Forward Motion

Forward

WORLD-CLASS RIDERS AND

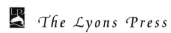
The Lyons Press

Motion

THE HORSES WHO CARRY THEM

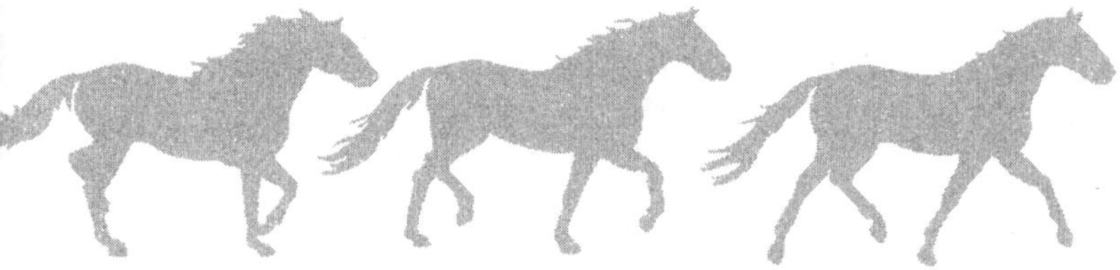

H O L L Y   M E N I N O

In memory of
Alton Huston and his remarkable Sandy

First Lyons Press edition, 1999

Book design by Abby Kagan
Printed in the United States of America
10 9 8 7 6 5 4 3 2 1

Library of Congress Cataloging-in-Publication Data
Menino, H. M.
    Forward motion: world-class riders and the horses who carry them /
    Holly Menino.
        p.   cm.
    Originally published: 1st ed. New York: North Point Press, c1996.
    ISBN 1-55821-692-8
        1. Taylor, Keith. 2. Gray, Lendon. 3. Kursinski, Anne.
4. Horsemen and horsewomen—United States—Biography. 5. Eventing
(Horsemanship)—United States. 6. Dressage—United States. 7. Show
jumping—United States. I. Title.
SF284.5 .M45   1999
798.2'0973—dc21   98-28429
                                                                    CIP

Contents

Keith Taylor and Fulmer Faktor (Christine Tchir)

Endurance and Magic

Play Me Right is dark, the color called bay, a black mane and tail on a brown horse. In his case, the brown is as dark as a mink, making him nearly black. He is quite plain. There is no white on him, and his only marking is the mulelike tan shading around his muzzle. For a dark horse, he has light eyes. They have a gold cast, and they look out at the world, its creatures, its events, without wavering. I was sent into his stall to put on his halter and lead him out. Play Me Right glanced down at me briefly and endured the halter. He had seen me before, and I was not important, not worth his attention. He resumed his vigil, ears forward, eyes ahead, watching for Keith Taylor.

Play Me Right is an athlete who competes in a sport called combined training, and Keith Taylor is his trainer. He is the horse's pilot, his guide, and his caretaker. I had visited them off and on for three years, and Play Me Right is so steady, so accustomed to the routines of all kinds of humans—the ones who pick up his feet and pound on his shoes, the ones who carry needles, the ones who drive up hastily and slam car doors, and the ones who just want to pat his hard-muscled neck—that it took quite a while for me to see how intently the horse watches Keith.

Keith had just gone around the corner to get something, and in a couple of minutes he was back. The horse's eyes stayed on him. The man is probably interesting for a horse to observe. He speaks very little, but he's always quietly in motion, a slight person with olive skin and dark hair who would have no trouble making jockey weight. Because of Keith's size and the sudden, bright devilment when he smiles, it might be tempting to think of him as childlike. But his remarkable focus and gutsiness quickly overtake that first impression. Also, he is very strong. I outweigh him by ten pounds but he is much stronger than I could ever be.

Keith works out of an old Pennsylvania Dutch bank barn near Pottstown, Pennsylvania. The lower story has been cut under the main floor of the barn into the side of a hill. This is where the small stable is. There are five stalls on dirt, and a brick floor runs between them. The place is very clean. The bricks are swept after each horse takes his turn being groomed, ridden, and returned to his stall. You can smell the horses, the leather and soap, and the earth behind the back wall, but a strong medicinal smell overpowers all these. There is always a bucket of warm water that has been laced with liniment. It is left there to be sponged on each horse when he has finished his work for the day.

I was there to watch the bay horse's work. I came back every few months because I was interested in the partnerships between Keith and his horses. I had kept horses myself for most of my life. But you can spend a lot of time with these animals, you can spend a lot of money, and still not know much. I had spent a good deal of time and too much money trying to know them, but my own case was hardly extreme. Some of my friends pursued their work with horses much more seriously than I did. Eventually they had to make sacrifices I would never consider, giving up jobs, marriages, family, and any possibility of financial security in order to continue with horses. I watched them making their choices, and I began to wonder what it was about riding and training that sustained them and what it was that engaged their horses. Communication between the horse and rider intensifies in athletic competition, and I began to wonder what training at the highest levels of sport was like, what it takes to make a good equine athlete.

I could have visited trainers who work with horses that race or steeplechase or play polo. There are many great horses in these sports, but they work in groups with other horses. Their bonds with trainers are not so highly developed, and the skills they use are not terribly intricate. If I had watched trainers working with Western horses that compete in cattle cutting and reining and roping, I would have witnessed processes that each have a momentum that leads the horse to increasingly sophisticated work with humans. These horses are fit for complex tasks, they develop strong individual bonds with their trainers, and in their work they must use a degree of independent judgment. There are many horses that excel in the speed, agility, and strength demanded by Western horsemanship—many canny ones, many generous horses. But the world of horses and riders is diverse, and I went to the corner of it I knew and loved best. I approached trainers in the three Olympic disciplines—combined training, dressage, and show jumping. Keith was the person I visited first. This happened by chance, but it provided a framework for my experiences with Lendon Gray, a dressage rider, and Anne Kursinski, a show jumping trainer. Keith's sport, combined training, is a triathlon that incorporates a phase of dressage and a stadium jumping phase, which is essentially show jumping, and then adds a third phase called endurance.

Play Me Right must maintain his competitiveness in all three phases. He is a Thoroughbred, the breed the Jockey Club registers to run at the track, but many Thoroughbreds never make it to the track. They are valued for their speed and athleticism, their sensitivity, and in the case of many horses like Play Me Right, their independence. So they're trained for other kinds of work, or perhaps they try out at the track, aren't fast enough to earn their keep there, and are sold into the hands of other kinds of trainers.

Play Me Right is small for a Thoroughbred, but he is fast and catlike. He also has a highly prized quality called boldness. It eludes description, but boldness is gameness and integrity combined with great confidence around humans. Boldness offsets a horse's instincts as an animal of prey, and Play Me Right's boldness was evident in the way he presented himself to me. He was grazing on the hillside with two pasturemates when he spotted me down in the hollow near Keith's riding arena. He lifted his head, then left the little herd to investigate. He stepped up to me, sniffed me crotch to boot to armpit, and when I spoke to him, he looked straight up into my face. He is not worried about humans or what they might put in front of him. He is the best horse Keith has had, and he has brought his rider to the highest levels of the sport.

On my first visit, he and Keith were getting ready for the Fair Hill International, one of just two combined training events in this country rated on an international scale of difficulty at three stars. Fair Hill would be Keith's first outing at this level. He strapped protective foam boots on the horse's ankles before riding him for me. A horse's legs are the most vulnerable parts of its anatomy, and with the Fair Hill three-day just a couple of weeks away, Keith was taking no chances with Play Me Right's underpinnings. The horse was being seen by a veterinarian and treated by an equine chiropractor. His shoes were being checked and refit by the farrier, and instead of galloping to stay fit, Play Me Right was loaded on a trailer and hauled to a facility where he worked on an underwater treadmill. This exercised his heart and lungs without the concussion of galloping. Today Keith would break training by actually riding the horse. "I'll be more relaxed in two weeks."

But he seemed very relaxed to me. He rode Play Me Right down into a field behind the stable, which dips between two old, irregular hills that characterize the landscape west of Philadelphia. The pasture, owned by Keith's parents, Millie and Keith Sr., is what development has left of an originally much larger farm, but it has enough room, just enough, to gallop. Behind the horse and rider, steam rises from the huge twin towers of the Limerick atomic power plant, giving the figures an unearthly backdrop.

They began slowly because even experienced horses begin work slowly. Keith is a quiet rider. No jerks, no jounces, no extraneous motion to distract the horse, which is probably a good thing, because once he permitted Play Me Right to begin trotting, the horse came to life, high-blowing, snorting in time with his gait, expressing his enthusiasm and his hope to be allowed to move faster. Play Me likes his work. He wants more of it. He wants to cut to the chase. Even in motion, covering ground at top speed, Play Me Right does not give the impression of a lot of horse power. He is light, catty. He has a way of making his efforts look easy. When he's jumping, he leaves the ground without grunting or audibly checking his breathing. When he's galloping, there is no thunder of hoofs, just the rhythm of his blowing. They skim the border of the pasture, picking up speed, flying over a giant log, moving even faster, like movie horses when the film speed is advanced. At a distance, their lightness has the quality of magic, something spontaneous flowing between man and horse, and they make me think of Walter Farley's descriptions of Alec's mysterious influence over the feral Black Stallion.

As a child I read all the Black Stallion stories, all Farley's Island Stallion stories, Marguerite Henry's *King of the Wind* and *Justin Morgan*, Mary O'Hara's *My Friend Flicka* and *Thunderhead*, and Enid Bagnold's *National Velvet*. From the time I was eight until I reached fourteen, literature took one form, the horse story. In the horse story, the hero is a boy—although National Velvet is a girl who disguises herself as a boy in order to ride the course at Aintree in the Grand National—who has a magical touch, and the horse is an animal whose speed and power depend on this touch. With their combined prowess, the horse and his hero overcome tremendous obstacles to reach victory. I realize now that the parable points to certain truths: every horse has a tale to tell—"What's his story?" is the first question many prospective buyers ask because the tale will have a bearing on how the horse responds to humans— and it does take a special touch to work successfully with horses. But when I began visiting Keith and Lendon Gray and Anne Kursinski, I wanted to get the real horse story, to find out what really goes on between the horse and the hero.

Watching Keith and Play Me Right more closely, I could see the carefully wrought physical responses that give the impression of magic. Keith had been working with Play Me for three years to build the fitness and skills that created this impression, and although his partnership with the horse was as complete as any long marriage, their daily work was still work for both of them.

The apparent harmony between Keith and Play Me was interrupted only once. Just ahead of a practice fence, a stride before takeoff, the horse raised his head against the reins and opened his jaws. He was struggling against Keith's control. The jump that followed was awkward. Play Me did not like waiting to take off. He wanted to stand back and fly at the fence from much farther away.

Keith looked back with a quick smile. "It used to be real bad, but now we have a kind of agreement. It goes like this." He presented the horse to the same jump again, and the flight over it was smooth. Immediately after the fence, Keith slowed the horse and dropped the reins. Play Me enjoyed the free rein. Given his head, he continued to trot around the riding arena, nose dropped like a dog seeking out a scent, and he meandered among the jump standards on a self-guided tour. Then it was Keith's turn to be in control again. He took up the reins and put the horse to a sequence of fences in which he was the one to determine the timing. Even at his high level of training, Play

Me was reluctant to give over control. He was convinced he knew the game better than Keith, and Keith was constantly, tactfully, negotiating—demanding, asserting one moment, then instantaneously humoring the animal.

The matter of timing is important. The cross-country obstacles in the endurance phase and the fences in stadium jumping come up very fast. There's not much time to negotiate. The clock will punish. What the rider is attempting is like setting out from the top of a ski jump on the back of the skier, and although Play Me Right is very talented, he is not an easy ride.

Strong convictions were not likely to be a serious handicap for Play Me Right, since they often go along with talent and speed, and a rider as sensitive as Keith learns how to accommodate and even exploit these characteristics. The two things that were more likely to work against him were his size and age. He stood five foot three at the shoulder, several inches shorter than most of the horses that win consistently, and at fifteen he had passed peak athletic potential. Like many aging competitors, he was haunted by old injuries. Two weeks before the Fair Hill event, there were signs of pain or stiffness in the horse's hindquarters. If it persisted, if there was any hint of gimpiness or anything arrhythmic in the horse's gait, the panel of judges and veterinarians would bar Play Me Right from competing.

The physical demands on the combined training or "event" horse are both grueling and exacting. In the first phase, dressage, the horse performs a set of formal school figures to demonstrate the purity of his movement and the correctness of his training. Then come the speed and distance and danger of the endurance test, and finally those horses that have been able to complete the first two phases are tested over a stadium jumping course that demands utmost agility and maneuverability. Top physical conditioning is required for the horse to be allowed to even attempt the three phases of competition. If, at any of the periodic veterinary checks, the horse appears anything less than 100 percent right and ready, he must be withdrawn from competition. It's a punishing sport, even in the way it is scored. The only points awarded are penalties: for imperfection of movement, lack of speed, lack of agility.

Even if Play Me survived the first veterinary examination and the jog-up after the dressage phases, he could still be "spun" from the vet box during the endurance phase or, even more frustrating, just before the final stadium jumping round. Keith wasn't talking much about these possibilities, because he never talks much. He's not shy, not unfriendly, just quiet—and unpretentious, which makes him well suited for a sport that routinely crushes the hopes of even Olympic veterans. When I ask him to explain something he doesn't elaborate. He just repeats the same words more slowly. Words mean what they mean, and adding more words cannot increase understanding. He hopped off Play Me, pulled away the horse's galloping boots, and said, "All I'm worried about is getting him there."

You can see competition in combined training as a three-day event or as "horse trials," which include an abbreviated version of the endurance phase. This briefer test is ten to sixteen miles shorter than the same phase in a three-day event. It drops two sections of "roads and tracks," which are ridden at a fast trot, and a steeplechase section, two miles over brush fence hurdles ridden at a flat-out gallop. Usually horse trials take place over a day or two, and they're like a three-ring circus. At any given time, you might see competition in dressage and horses flying over cross-country fences. I had helped organize horse trials, built fences, scored, and even owned horses who had run in trials. But until I went to Fair Hill, I had never seen a three-day event, where each day is given over to a phase and where during each phase you can compare every horse in the running.

Because the Fair Hill three-day is internationally ranked, the event is governed by the Fédération Equestre Internationale, as are the Olympics, the World Cup competitions, and the World Equestrian Games. U.S. representation on the FEI is managed by the American Horse Shows Association, which issues rules for combined training but leaves most practical aspects of the sport to the U.S. Combined Training Association. To ensure that the horses are able to even attempt the fences and speeds, they must go through a graduated process of qualification through the Training, Preliminary, Intermediate, and Advanced divisions; the horses are graded on a national scale, with points being awarded for placings according to the difficulty of the division.

It's an elite and fragile universe of horses that succeed to Fair Hill. This year an important international event in Holland had been canceled because of a long spell of heavy rains, and a number of riders from the Dutch and French teams elected to ship their horses across the Atlantic to compete at Fair Hill. Because of the large entry, the dressage phase took place over two days. I arrived in the bright sunshine of the second day to find that Keith had ridden his dressage test the day before. Even though it was a Friday and a working day, a couple of hundred spectators parked their vehicles—an egalitarian mix of Mercedeses, pickup trucks, and road-weary compacts—in the field across the road and crowded into portable bleachers alongside the dressage ring, where the riders performed in formal attire —tails and top hats—and the horses were turned out glistening, with their manes knotted tight against their necks in traditional braids.

It was the last weekend in October, bright, breezy, and unusually warm. In the woods behind the dressage ring, the maples stood out with hot oranges and reds against the duller oak leaves. The people in the stands around us and established in folding chairs along the crowd-control tape wore sensible clothes, good sweaters and tweeds, oilskins, jeans and sweatshirts, sturdy shoes and barn boots flecked with straw. The only dress pumps in sight were on official feet. Our group of spectators spanned the generations— people in their twenties, forties, sixties, and even eighties. I am always impressed by the number of elderly people who turn out to watch the important combined training events, because the sport requires active spectators. While you can sit down to watch the dressage rides, after that you're up and out on the course, walking and standing for two days. But I think where horses have been a concentrated pastime, such as the Philadelphia suburbs only an hour north of us, many of the old people come out because they remember great horses or maybe they owned just one good one—and the sight of a good one is something you develop a hunger for. Many of us had driven long distances to be there in time for the first ride, and some had responsibility for a horse in competition. Most have been up for hours, and this accounts for one of the joys of watching horse sports: you can eat a chili dog at nine in the morning and no one will glance twice at the beans and tomato sauce.

A dressage audience is what a tennis crowd used to be, a quiet gathering with an intense focus. There is protocol to be followed: you keep your dog quiet, you do not talk during a ride, you hold on to any paper items that might blow across a horse's path, and you do not get up or leave your seat until the current horse and rider are exiting the arena. You do not take the horse's attention from his rider.

Although the movements in a dressage test for combined training are not so difficult as those that would be
asked of a horse that specializes in high-level dressage, they
are challenging enough so that the horse must concentrate
and work in instantaneous cooperation with the rider. For
event horses, who are conditioned for peak speed and stamina, the dressage phase can be the toughest. They are too
fit, too hot, to have much patience for the detailed accuracy a good test requires, and many of their tests seem
perfunctory and hasty. Nevertheless, dressage can make or
break the score for the whole competition, and as combined training becomes more highly competitive, events
are more frequently won on the basis of a good dressage
score.

Keith's score was already posted. The penalties were
fairly high. I wasn't surprised. Keith had told me Play Me
Right did not usually do well in dressage. "You know, he's
a kind of common mover, and dressage is tough for him
because it isn't what he *likes* to do." For Keith and Play
Me Right the dressage test was the price of admission to
the start of the endurance phase.

We left the scoreboards and went out to the woods to
walk the cross-country course, which wound in and out
under the trees' highly colored canopy. In the endurance
phase, the clock ticks, the miles stretch over natural ter-
rain, and in the last minutes—after two sections of fast
trotting for several miles and the two-mile steeplechase—
the horses come to the cross-country course. Touring the

course before the rides is one of the great pleasures of spectatorship, which is fortunate because it's also a necessity. The horses run the course individually, and there is no way to see a horse at all the obstacles. Traveling the course on foot and confronting the fences as if you were riding up to them is the best way to understand the action as you hear it announced. Often you walk in the company of competitors who are sizing up the technical problems at each fence, rehearsing the approach, counting strides, and revising the angles they plan for their approaches. It's not unusual to come upon a rider standing off from a fence, just standing, thinking about whether the horse will require a slower way through the problem and how to make up the time later.

My hometown experience with cross-country fences had not prepared me for the obstacles on the course at Fair Hill. They seemed huge—not just in terms of their height or breadth or depth, but in terms of the technical problems they presented—and they made me acutely aware of what is so often said about these obstacles: they don't come down. They are called fences, but they're made of heavy timbers, stone, and earth, and the ditches are real. They are deep, wide spaces that interrupt solid ground. This course, like most, was laid out with a momentum in which each fence builds progressively on the skills demanded by the fence ahead of it, until the last section, where the horse's fatigue is usually acknowledged and the footwork necessary to negotiate the obstacles becomes less tricky.

About three miles out on the Fair Hill course, we came to an obstacle that seems emblematic of the demands of combined training. The horses are more than ten miles from the start of the endurance phase when they approach the Serpent. With nearly an hour of exertion behind them, they gallop up to a wood panel poised on the edge of a gully. Before they leap, this is all the horses and riders see, the panel and the tops of the trees on the opposite side of the ravine. They land on a narrow shelf above a creek, jump down to the creek bottom, spin to the left to lift off over two more consecutive wood panels, and scramble up the other side of the gully to jump out over a last panel that opens on more galloping, uphill. Meanwhile, the clock is merciless, running ahead. I had never seen a fence this treacherous-looking, and when I looked down into the Serpent from the panel at the top of the gully, I wondered how a horse and rider would get out alive. Because it had been built by experts, I understood that it must be feasible to negotiate it, but I wasn't sure I wanted to watch Keith and Play Me drop over the panel and out of sight. There were other hazards on course as well: the Waterloo Rails, a pen of heavy timber built on a sharp downhill slope, where the horses galloped down to the first wall, dropped in, and immediately sprang out, and the Chesapeake Crossing, a big pond the horses landed in after sailing up over a huge log. Even if you are a seasoned combined training fan, fences like these are enough to raise your pulse. I wondered about the reactions of the riders who were in-specting the course along with us. But Keith had said about Play Me Right, "There's nothing out there that's going to surprise this horse. He's already jumped every kind of cross-country fence they make. So it's just a matter of getting there."

On our way through the last few fences, we heard the officials paging Keith Taylor, and I worried that there was a problem. But with the stable area closed under FEI rules there was no way to find out until the next day, when we would see which horses actually started the endurance phase. Heavy rain began before dawn. Umbrellas congre-gated near the fences, and the television crews covered their equipment with plastic bags requisitioned from the concession stands. Water ran off the roof that sheltered the big boards where the scores were posted. Many horses had been withdrawn after dressage, including Play Me Right. "There was *some*thing wrong," Keith told me. "He could have made it around the first three sections of the endurance, but I wasn't sure about the cross-country." Keith consulted the official panel of veterinarians, who watched Play Me Right trot beside Keith on hard pave-ment. None of them was certain the horse was moving normally and free of pain. Although they weren't sure there was a problem and couldn't begin to pinpoint the source of any pain, none of them could offer any assurances either. Pushing the horse beyond his capacities could have resulted in falls or strains that could have impaired him permanently. Keith decided to retire. "So frustrating, but . . . discretionary."

That first year he did not get to run the fences at Fair Hill. He stayed long enough to help take care of friends' horses as they steamed into the vet box before the start of the cross-country obstacles. Then he loaded up Play Me Right and drove home to Pottstown. The horse was at least five years beyond his prime, and he had an unspecified weakness in his hindquarters. Because horse locomotion is rear-end drive, this was Keith's most serious concern. Every galloping stride, every jump, requires a horse to push off from behind. Keith didn't know if Play Me would ever be able to run another cross-country course, much less an international-level course. "We'll see," he said. "I'll just have to wait . . . and see." Later when I learned how the dark bay horse had come to him, I realized how emotionally difficult it was to pull Play Me out of the lineup. But Keith was not talking about what was hard, and when I made sympathetic noises, he deflected them. "You know," he pointed out, "I still have Faktor"—a young horse who had recently come to him for training—"and he goes next weekend in Virginia."

The combined training season is a series of weekends strung out at intervals over the months when good weather is possible. The one-day horse trials and three-day events

begin in Georgia and the Carolinas on early spring weekends. The season travels north with summer into New England and Canada, and then moves south again to end in
November in Virginia. While every town in this country
may have a ball diamond, there are only a few sites for
combined training, and event riders become accustomed to
traveling four and five hundred miles to compete.

There's precious little money in the sport. Only a
handful of events have begun to offer prize money, and
these sums are laughable in comparison to the corporate-
backed purses offered for world-class show jumping and
dressage. The trainers who make a career in combined
training get their livelihood from teaching people and
horses how to do what they do. Then on the weekends
they test out the skills they've been developing in their
horses and students. Like many trainers, Keith trailers the
horses himself, and because horses can get restless standing
unattended on a trailer, this means no stopping and a lot
of fast food. Once on the grounds, he is trainer, jockey,
groom, water boy, and instructor. He and his horses and
pupils meet their tests, and they excel or they fail, they're
in the ribbons or they're turned away from the competition
at the vet check. Then it's a long drive back to the Pennsylvania Dutch barn and the work that is ongoing there,
training.

He likes doing things by himself. Difficult as his financial situation sometimes becomes, he treasures his independence and speaks with a kind of dread of the possibility

that he could take a job training horses for a private stable and have to negotiate over goals and methods. He's young—twenty-seven when I first visited him—and newly professional. He has the energy, physical resilience, lack of family responsibilities, and talent to allow him to keep up his strenuous routine until the right horse comes along to change his fortunes. For the time being, it works. He lives with his parents on the tiny farm—"they've been very gracious"—pays no rent, and stables his competition horses board-free. Other bills pile up—veterinary, chiropractor, farrier, physical therapist, entry fees, travel—and to pay these Keith shuttles from stable to stable in the West Chester hunt country near Philadelphia. He teaches, mostly adults, and he rides young horses, problem horses, and horses ready for more advanced riding than their owners can give. He's a riding master, he's a cowboy, he's a jockey. "I try to do six jobs a day," he said to explain how he manages to make ends meet. "Doesn't always happen, but six jobs is what it takes." Then after eight to ten hours of riding for other people, he can train his own horses.

Shortly after the disappointment of Fair Hill, I caught up with Keith to ride along with him. He was in an optimistic frame of mind—Play Me seemed to be returning to soundness, and next season was still a possibility. We set off on his rounds in his aging Saab. Keith has calculated to the minute how long it takes to go from one stable to the next. He gets off a horse in time to put him carefully back in the stall, throws his saddle, pad, and helmet in the backseat of the Saab, starts it, and accelerates out of the drive. His route is mostly narrow twisting roads through posh suburbs, but their limitations are not evident to Keith. He's intent on getting to the next job at the appointed time, mindless of his speed and of the curves or ungoverned intersections, and it's clear from the rattles and creaks in the Saab that the car has not received the care that Keith's horses get. Our time in the Saab should be a good time to talk, but I'm often holding my breath. We are never late to an appointment, and Keith goes calmly to the horse or student that waits, unflustered by the speed and close calls that brought him there.

He rides the horses other people can't deal with, the spookers, the rearers, the runaways. He rides very young horses that haven't often had a human on their backs. That fall, he had a rearer to deal with. This category of animal is despised and feared by horse people. When a horse resists his rider by standing on his hind legs, it takes only a small error by the person on his back—a slight shift of weight, a tightening of the grip—to pull the horse over backward, bringing a half ton of bone and muscle down on the rider. Even experienced trainers will send a rearer away—someplace where another rider will take more risk or be rougher than they will.

One of the first horses I had was a rearer. He was a tall chocolate-colored gelding named The Candy Kid. My high school friends and I never bothered with saddles. We hopped up and rode the back roads together. In the company of other horses, my horse was never much trouble. But asked to walk into the world without company, The Candy Kid found many details of rural Ohio scenery that gave him pause. Soon pausing became stopping, and stopping became rearing. The first time he went over backward, it must have been because I pulled him over. I started scrambling before we met the road, so that he only pinned one of my legs and only briefly. After that first time, he began to anticipate just how he would manage this without hurting himself, and he stood up prepared to flip over. I was more afraid of having the horse taken away from me than I was of slithering off a swooning horse. So I kept the problem quiet for a long time, hoping the horse would forget the trick. Of course, he never did, and now I look back on the chocolate-colored horse with real alarm.

"Do you think the horse is really worth the risk?" I asked Keith about the rearer he had been hired to ride. He said, "You haven't seen him yet. He's *beautiful*"—a word that for him has as much to do with ability as with appearance.

"So what's the cure?"

"Keep him going forward. That's what everybody says—he can't go up if he's moving ahead."

At the close of the morning, we were headed back to Keith's place, hurtling around corners because there was just enough time for Keith to ride his young horse for me and still get to his first afternoon appointment.

I said, "What happens if you break your neck on one of these animals?"

"I can't get disability insurance, if that's what you mean. If I get hurt, I don't get paid *and*"—the worst possible scenario—"my own horses don't get ridden."

We wheeled into the Taylors' drive, and he was out of the car instantly, going to catch up the horse. But I was still worrying the question of risk. I wondered if a person could get to a certain point and lose his taste for risk. Keith was leading out a massive red horse and snapping him into the cross-ties. He was focused entirely on the big horse, but he listened in a disinterested way to my question. "Well," he said, "this is my life."

He would say this fairly often to explain himself or some course of action: "This is my life," a self-contained rationale. In fact, there never seemed to be any other life for him. His first contact with horses came before he was school age. His parents bought a gray horse for his sister, who joined the local Pony Club. U.S. Pony Clubs are roughly equivalent to Scout troops for horse-crazy kids, and they have been the training ground for many Olympic riders and other professionals. Keith tagged along to meetings and rode the gray horse as often as his sister—"I was better at it than she was." But he was very young and small even for his age, and Millie Taylor worried because the horse was too large for her son to manage safely. Keith refused to recognize his limitations, so eventually a chubby palomino pony was purchased. Keith was allowed unrestricted access to the tiny horse. Millie watched out the kitchen window for a period of weeks while Keith fell off, chased the pony, climbed back up, and fell off again. But it wasn't far to fall, and he rode the pony every way imaginable—lying down, standing up, and for one long afternoon, clinging upside down to its belly. The pair bonded like two in the same species. Millie returned home from work one afternoon and found Keith in the kitchen digging through the refrigerator. The pony was beside him, nosing about the vegetable compartment.

While other high school kids were struggling to discover who they were—drinking and cruising and smoking dope—Keith was riding school figures, shoveling manure, and pacing off distances between fences. He came up through the ranks of Pony Club and earned the coveted A rating. He now rode a Thoroughbred mare, and the pair was tough to beat in the jumping classes at local horse shows. Then Keith was introduced to combined training, and he started college. He competed his mare successfully enough for her to be named Mare of the Year by the U.S. Combined Training Association. Keith had a friend several years younger than he was. She had begun competing a dark Thoroughbred horse at the lower levels of eventing, and they sometimes rode together. When she was killed in an automobile accident, her parents couldn't bear to sell her horse. They turned Play Me Right over to Keith.

His life with horses exerted a strong pull, and he broke away from school for extended periods. It took a long time to finish. When the noted British dressage trainer Robert Hall offered a clinic nearby, Keith signed up. "He liked the way I rode. He liked to watch me ride." It wasn't long before Keith was taking off from school again, loading up his horses for South Carolina, where Hall had offered him a spot as a working student. Hall had become famous for his noncoercive, Zen-like methods: "It doesn't matter. Stop trying. Don't be careful."

In spite of the geographic distance that separates them, Keith still considers Hall his mentor, and the Hall philosophy was especially evident the morning we rushed back to his place so he could ride Faktor. Fulmer Faktor, who belongs to Hall, is the antithesis of Play Me Right. Where the bay horse is light and moves like a unicorn, Faktor is powerful and carries himself like a warhorse. Where Play Me Right is wise and experienced, Faktor hasn't got a clue yet. And while tragedy brought the bay horse to Keith, a marriage brought Faktor.

During the previous summer a wedding took Keith back to South Carolina, and on the trip down he obliged a friend by trailering a broodmare. When Keith had unloaded the mare, Hall looked at the trailer and remarked that it would be a waste to drive it home empty. Why didn't Keith take a training project with him?

Even though a trainer may have a good horse working at the top of his sport, the horse's time at the top will be limited, so competition riders are always on the alert for younger talent to bring along. Keith knew which horse he wanted, but Hall wasn't prepared to turn over that particular horse. Negotiations continued during the wedding festivities, and eventually Keith loaded up the horse of his choice.

Faktor is a red horse, not copper-colored like many chestnuts but red like a kidney bean, and quite big. When he stands in Keith's stable, his head comes within an inch of the ceiling. He is heavy, large-boned, one of the European warmbloods that result from crosses of cold-blooded draft animals on hotter Thoroughbreds.

When Keith worked for Hall, Faktor was a young stallion and a bad boy with some rank habits. The most troublesome trick in his repertoire was his ability to get loose. He had figured out how to escape from a handler by standing on his hind legs and flinging a foreleg over the long stallion lead. He became quite accomplished at this maneuver and a real liability for Hall, who is responsible for an extensive breeding program and the welfare of many visiting mares. Accidental couplings are intolerable, so Faktor was not to enjoy being a loose cannon for long. He was castrated and became a gelding.

Now, the big red horse seemed very quiet, lamblike, and I saw only two signs of his naughty past. Faktor has very large round eyes, and they have a jokey look in them. It isn't an unkind or shifty expression, just a look that promises that some funny things can happen in his world.

Then, as he stood for grooming, Faktor waited until Keith turned away for a different brush to raise his foreleg and pound his hoof on the brick floor. Keith spoke to him sternly because pawing is considered aggressive behavior, an antecedent to striking and rearing. The word from Keith was satisfying to the horse. Faktor stopped striking the floor and looked pleased. He likes to keep Keith engaged.

"I rode this horse when I was with Mr. Hall," Keith told me. "In fact, I think I was the first person to back him."

In other hands, a young horse might have shown signs of anticipation or nervousness as he was being readied for work. But Faktor stood like a riding school horse as Keith mounted, hopping up from the right side just as easily as if it was the orthodox side of mounting—I admired this little feat and have tried it now several times, but it seems impossible, like writing with my left hand. Keith settles instantly, and when he sits, every part of his body adapts instantaneously to its use on the horse. He is more subtle with his legs and back and whip than most riders I've seen, and he has hands that float the reins with the horse's head.

"This is probably going to be pretty boring," Keith warned me. But what I saw was hardly boring. It was the beginning of the very long process of training a horse. I caught a glimpse of the initiation of skills that take a horse as far as Fair Hill. Everything was new for Faktor, from the moment he ambled into the fenced riding arena. After a long warm-up at the walk, Faktor began to trot, and it was hard for me to detect Keith's physical give-and-take with the animal. It looked like sleight of hand. But Keith was negotiating, and it was work. After a few minutes, he shed his jacket.

The red horse's ears were forward. His neck rounded downward and he leaned ahead. Keith's hands moved with the head, not changing anything about the rein, and the trot was bigger, more purposeful. Off and on, Faktor's head jerked up resentfully. His eyes bulged a little. He was anxious about this way of moving, or at least he was questioning. The jokiness was gone. There was no humor in his face, but he continued to trot in this strange new way, flipping his head occasionally, until his trainer rode him down to a walk and gave him a few loud claps on his neck. Faktor had done what was asked, to move forward. To my eyes, the big horse was trotting in a pleasant rhythmic way, then suddenly—without changing speed—he moved forward with power and intent. Keith did something, made some change, but I couldn't catch it.

"What are you doing there?"

"I'm asking him to reach out to me."

The word "ask" is used constantly by trainers and riding instructors. "Ask the horse to canter," "Don't ask him to jump so soon." There seems to be an underlying assumption that language is the basis of training transactions. Although words do cue many training operations, the real language at work has more to do with bodies and movement, balance and physical signals.

Faktor was just beginning in the language that produces an event horse, and there were many basic responses he had yet to learn. He was only a very recent initiate to jumping, and his lesson with Keith began by trotting through a row of poles spaced out on the ground. When the horse handled these smoothly, Keith presented him to a low fence. Faktor trotted toward it, looking very hard at the rails ahead of him. His eyes bugged, but he did not hesitate. He fixed on the fence as he gathered himself, then heaved himself up in a very big effort and cleared the top rail by more than a foot. This is typical of a horse who is green and hasn't learned to match his effort to the size of the fence. Faktor was not confident, but he did what was asked, and he flipped his head only after the fence. Keith gave him several solid claps of approval on his neck and turned him back to the fence for another go at it. When Faktor began to move forward over the single low fence with some ease, Keith put him to two, then three, then a series of four. The horse was finding out how to see heights for what they are, to judge distances in relation to his own body size. Faktor was gaining fluency in the discourse of height and distance. More claps on the neck. Faktor relaxed at once. His head dropped and he doodled along at a walk, mischief and fun returning to his expression.

The big red horse was learning. He was engaging in the ask-and-answer of training. Responding to his trainer by making a new physical effort was the beginning of a different kind of connection with humans, a connectedness of mind and activity. Faktor was evidently taking pleasure from that connection, and as I watched this, I realized how intimate the process is. It is not readily shared.

We stripped off the saddle and began walking back up the hill to the stable. The horse stopped and lifted his head. There was a wet patch across his back the shape of a saddle. A warm vapor rose from his neck and shoulders and haunches, and behind him the towers of Limerick gave off two columns of white steam. Faktor has a heavy outline, like the bronze horses ridden by local generals in park statuary. But he is not a horse from history or fable. He is a contemporary horse, an animal living among industrialized people with their machines and roads and constant building. His ears and eyes focused urgently on something beyond the road at the top of the hill. He was hearing something. Keith and I heard nothing. But whatever Faktor heard worried him.

I wondered if he lacked courage. I remembered the bay horse Play Me Right lifting off to jump, and I had difficulty imagining the large Faktor in that confident, effortless flight. I couldn't seem to see him in the subtle transitions of dressage. At this point, the horse seemed to labor at everything he was asked to do. I knew Keith was quite taken with Faktor, but I wasn't sure I liked him.

"Keith, what do you think of this horse?"

"I like him."

"I mean, how *good* is he?"

"He's good."

"But where do you think he will *go?*"

"He could go to the top—at least in dressage or in show jumping. I don't know about combined training. I don't know whether he'll have the speed. Speed is the big question. But we're going to find out about that."

I asked when Keith would have a sure sense of where the red horse's abilities lay and where the horse would go in life. He looked amused at my emphasis on schedule and outcomes and certainty, and he gave Faktor a last clap on the neck. "Sooner or later."

Keith Taylor has a special touch with a horse, and if he had been a hero in one of the horse-and-boy fables, that's all he would have needed. Walter Farley's young man Alec had the magic touch, and all he needed to win the climactic race was the remarkable Black. But Keith is a real-world rider working with horses as they really are and with the business realities of making a living with horses. He was at the beginning of this process when I met him and his horses. By the time I met Lendon Gray, she had already survived some long, tight times before interest in the sport of dressage developed enough to support a teaching operation, and when I began to follow Anne Kursinski's riding, she had worked her way far enough up the protégé and patronage systems of show jumping to have a relatively secure enterprise.

Keith went into the winter looking forward to running Play Me Right in the next three-star international event. He had a lot going for him: he was a strong talent with good training and a growing reputation; he had one good advanced-level horse that just might have another year in him and a big, strong, promising young horse. But the plots of most real-world horse stories turn on money, and this was the one force working against Keith.

Bad weather closed in, and unlike better-established trainers, Keith had no indoor riding arena where he could exercise the horses. A distant neighbor with an indoor facility allowed Keith to work there, and this meant riding each horse along the edge of the icy roads before work could even begin. Through the early part of the winter he managed this way.

Keith's "jobs" fell off. His students stopped riding for the winter, and he kept up work with only those few whose horses were housed at indoor facilities. His resources dwindled. He told me he was thinking of applying for a clerk's job at the local convenience store. "I almost went under," he said in January. In February, storms covered the East Coast from western Pennsylvania to Virginia with a thick coat of ice. It crusted the ground for several weeks, and since the footing was too dangerous for horses, they were confined to their stalls. Play Me Right seemed to have recovered from the problems of Fair Hill, and Keith continued to exercise the two horses along the edge of the road. Even this was risky. "They both spook easily," he told me, "and Faktor really scoots. He's pretty strong to stop." Once Keith found himself suddenly riding through a neighbor's young wheat when only a moment before, the horse had been walking quietly along the road.

Keith was aiming for the Kentucky Rolex with Play Me Right, and for Faktor he had in mind May trials in Virginia and June trials in Massachusetts. He arranged to stable the horses temporarily at an indoor facility. "But I've got to get them out of there in a couple of weeks," he worried. "It's so dark and not very natural—it's better to work them outside."

The ice began to melt away, and the rain started. Keith added a long oilskin outback coat over his winter gear. Faktor continued to look suspiciously at objects along the road, and once when the rain was heavy he shied at a puddle. Keith did not excuse the horse. He directed him back to the puddle and began to insist forcefully that the horse walk by it. "I was getting after him a little," he admitted. The red horse reared, which didn't surprise Keith, who came forward against the horse's neck. But as Faktor stood up, he indulged in a familiar habit. He flipped his head, driving the bony bump between his ears into Keith's nose. When his rider dropped away from him, Faktor turned and ran. Keith felt sharp pain and then nothing. He got to his feet and walked to the nearest main road. The rain was so heavy he didn't realize that most of the moisture rolling down the oilskin was blood. A truck driver saw him standing in rain and blood and pulled over.

Last Scene and Lendon Gray (Cheryl Bender)

The Ponies Are Talking

They have no syntax," the Stoics reasoned about animals, "therefore we may eat them." My friend Roger is less self-serving but shares this belief about animals and language. On a weekend trip to Saratoga, he went to the track but did not bet. "I'll step up to the window just as soon as the ponies start talking."

Lendon Gray knows the ponies are talking, and she knows how to talk pony. The two-time Olympic dressage rider and the people who train under her spend their days—or, at the very least, their weekday mornings—working out the syntax of classical horsemanship. This is the same horsemanship that Keith Taylor trains for in the first phase of combined training, and its movements are also the basis of a show jumper's skill. But at Lendon's

Gleneden Farm, dressage is practiced intensively as a discipline unto itself.

Lendon has a big, clear voice. It is pleasant and it is very loud. She may be riding a turn at one end of the huge indoor arena and giving instructions to a pupil riding along the other end, and when her voice reaches the student on the horse, I'm sure it is still loud. It seems remarkable to me that Lendon can speak at the same time she is riding, let alone analyze someone else's riding and call out instructions to improve it. She is putting out a good deal of physical effort, and this is something many riders I've talked with feel is generally unrecognized—"Yeah, all you gotta do is sit up there, right?" was the way one of Keith's event rider friends typified the general run of ignorance among sports fans. But Lendon is able to ride, watch her students, and articulate what they should do. It is impossible to ignore, but her voice always suspects that it is being ignored. "You have asked the question," Lendon declares, "and now you must have an answer. You *must* get an answer."

Fifi Clark, like most of Lendon's clients, is female. She is tall and in her fifties, silver and gold patrician. She is fit and determined and riding a hefty bay mare. It is ten o'clock in the morning, when traffic in the Gleneden arena near Bedford, New York, is at its peak. Just north of New York City only a few miles from the Connecticut border, Bedford, the village and the outlying homes, scatters judiciously among private grassy hills, old trees, and stone walls. You can't see many of the houses from the road.

Similarly, the house and barns on the farm where Gleneden is housed are set back from the road at a protective
distance. Lendon's operation shares the estate with a Thoroughbred breeding facility. The indoor school is the largest
of a group of yellow buildings set in the middle of shady,
fenced paddocks and protected from travelers on the
county road by a lane with a security gate and the big
maple trees that border the road. The cars parked in the
stable courtyard are Mercedeses, BMWs, and Volvos. Lendon has left her much-used Subaru just beyond the wide
door to the indoor arena and made a quick circuit through
her office in the front of that building before heading out
to the mounting block, where one of the Latino grooms
waits with a small gray horse. Five other women have
brought their horses into the building. There is a lot of
discussion going on, many attempts to talk pony.

    In its strict meaning, "dressage" is simply the training
of the horse. In use, though, dressage has become the art
and intensely competitive sport of training in the classical
movements. Like figure skating, dressage emphasizes the
power and beauty of motion, and competitors in the sport
are judged on accuracy and the élan with which they perform prescribed movements. The movements the horse and
rider are asked to perform depend on the horse's stage of
development. At the lower levels, the tests demonstrate
simple movements with long intervals of modulation between different gaits and maneuvers. The higher the test
level, the more difficult the coordination asked of the horse

and the more rapid the transitions between movements. Lendon Gray is one of the most successful practitioners of dressage in the United States. She has won more national titles than any other trainer in the country, and she has become known for her ability to make brilliant, expressive horses out of unlikely candidates. She has a round valentine face and a small, sweetly shaped mouth. She is forty-six, not large, but she carries herself with the same great authority carried by her voice. Sitting a horse, she looks taller. She is straight and supple, centered, and there is relaxed symmetry in her activities. When she enters the grand prix arena in formal attire, her concentration is fierce, a force that shuts out show grounds, audience, judges, everything except the white boundary of the arena, the letters that mark the points of transition, and the responses of the horse under her. This concentration is evident in her riding at home. Somehow, although she is answering my questions and supervising the other women riding in the arena, her focus never drifts from her horse, Last Scene. A small gray horse with black legs, he has delicate, refined ears and the round dark eyes you see in illustrations of ponies in children's books. He is the horse Lendon has most recently introduced to the movements of the grand prix test, the ultimate competitive goal of a dressage horse.

Big windows above the long sides of the arena allow daylight to fall on the riders' heads and the soft brown footing. Hoofs grind the tanbark into fine shreds and mix it with the animals' manure. Most of the women here are clients. They pay Lendon to stable their horses under her management and ride under her tutelage. As is typical of dressage riders, most of these students develop an intimate knowledge of their own particular horse. While they do not actually handle routine stable work, they devote many hours beyond their time in the saddle to their horses. They become keen observers of horse motion, always on the alert for some encroachment on their horses' soundness, and they put a good deal of thought into the details of the horses' well-being, feeding and fitness, turnout and blanketing. One longtime observer of horse sports remarked to me that a dressage rider could look at the horse standing in his stall and instantly detect a new fly bite on the animal.

Although some of Lendon's students compete in dressage, a number of them ride with her for the sole purpose of initiation into the art. They are the legatees of a school of horsemanship that dates from Renaissance Italy but was first put forth humanely as a "scientific" system in 1623 by Antoine de Pluvinel, who ran a finishing school for young noblemen and was the riding instructor of Louis XIII. His *Manège du Roy*, a philosophical dialogue about the horse and its proper education, guided many other riding masters across the Continent who were entrusted with the education of aristocrats. In the manège, the riding school, the activities of the horse were routinized as gaits and movements. He was taught to travel straight forward by "treading the ring," trotting or galloping along narrow paths, and to be "just" in his turns by repetition of various exercises.

In Lendon's arena also, each woman rides a deliberate pattern. Each horse keeps his own cadence, and the arena reverberates with muffled polyrhythms cut through by the voice of Lendon Gray. She works like a ballet master with dancers as they work on their exercises at the bar—exhorting them, calling out sharp corrections, tapping a leg here, reaching up to reposition a hand. Keith Taylor has progressed through many of these lessons with Play Me Right and through the less complex tasks with Faktor. Play Me Right, for instance, can canter sideways and forward at the same time, the legs on either side of his body reaching past the legs on the opposite in a lithe X—no small feat of coordination—and Faktor does well in the turns and bends and adjustments in speed that are expected of a horse in the more elementary dressage tests. By way of becoming competitive in combined training, Keith has had to become a rider who could compete in dressage at the middle levels of the sport. The skills Play Me Right develops for the dressage phase are essential to the balance, thrust, and precise footwork the horse will call upon in the cross-country and stadium jumping phases.

What is different about the movements Last Scene performs with Lendon is the degree to which his balance is shifted back into his hindquarters. He is building tremendous isometric strength in his back end, and this allows him to be so light off his front end that Lendon can control

an impressive range of motion with imperceptible movement of her seat and hands. When she gathers Last Scene
back into himself like this, he is "collected," and collection
is the legacy and mode of the manège. Beginning with the
likes of Pluvinel and his British counterpart, the Duke of
Newcastle, Renaissance trainers began to incorporate this
rebalancing of the horse into a system of training that
evolved in parallel with the very different techniques for
educating horses and riders for military operations in which
the rider leaned or perched forward on the horse. These
military methods appear to have their origins among the
nomadic warriors of Eastern Europe, for whom speed was
a far greater consideration than poise or precision. Tension
between the techniques of the manège and the traditions
of the cavalry and hunt field has been ongoing. It has produced lively theoretical debates and a good deal of cross-
fertilization among diverse schools of equitation, and it
continues to act as a dynamic link among contemporary
teachers and masters.

Dressage, the art for art's sake school of horsemanship,
does not involve the speed or danger of other horse sports,
but many top riders have left fast times and big fences for
the intellectual stimulation of manège and the rigor of its
search for purity of movement. Among Lendon's students,
there are artists, cooks, lawyers, financiers, and an actor.
They are, on the whole, people who understand the re-
wards of submitting to a discipline. As practiced at Gleneden, dressage is an ongoing process of development in

which competition is an optional test of progress. It is a continuum of momentary performances as transitory as passages of music, and each of these performances is the result of the rider's intimate communication with the horse.

The big stolid bay mare carrying Fifi Clark is thundering along pleasantly. Everything about this mare is large, weighty. She is heavy equipment going to work. But there is something about the way the horse is responding that Lendon does not like, and it sends her voice up a few more decibels. "She is *ignoring* you. *Demand* an answer."

The silence Lendon hears is the curve of the mare's body as it trots a circle. She will not accept it. Although she continues to engage in dialogue with her own small gray horse, she keeps after Fifi until something shifts briefly. The change is momentary, but it is the answer Lendon was listening for.

"Good, Fifi—*super!*"

The language Lendon's students struggle with is unuttered. Its grammar is split-second sequencing of physical cues—the insistence of the legs, the tightening of fingers, the pressure of rein on bit and bit on mouth, the straightening of the rider's back, the sinking of her loins. This is the nature of riding because it is the nature of the horse. Horses do make sounds, and we have the words "neigh" and "whinny" to label them. But horse noise is usually an uncontemplated call, and what prompts it is some kind of need, such as hunger or, because horses are intensely social and their society depends on physical proximity, separation from other horses. But even when the horse is communicating with others in his herd, noise is cruder expression than what the horse achieves with his body. Dominance and fear, love and elation, are explained by one horse to others in terms of motion and place—where a horse puts her body, how he carries his tail, the curve of her neck, the angle of his ears.

Movement is the basis of our communication with horses, from trivial conversation to artistic endeavor. It conveys the subtlety of horse thought and horse intention. Lendon doesn't talk about *language*. She doesn't tell her students they are learning a language or insist they conjugate the verbs correctly as she did when she was tutoring Latin and Greek at Sweet Briar College. She assumes the language is there, and she assumes that her students are just looking for the right vocabulary. Later, when I began to visit Anne Kursinski, I found that she makes the existence of this language explicit in her teaching and writing about show jumping. Once a dressage rider herself, Anne still talks pony, and in her book on technique and training, she advises all who would ride to be aware that that is what they are doing, using language. When you sit behind the horse's shoulder and wrap your legs around his sides, you are in the conversation pit, the physical center of the animal. The sides of some horses vibrate, making a connection of your nerves with theirs. The sides of others seem dead, not like living matter but like a mattress. Either way, every impulse of the horse will pass through you, every word the horse says—if you can only hear.

Lendon listens to all kinds of horses. She is a democrat. In the dressage world, where big horses with extravagant movement have created the standard, this makes her an iconoclast. She rides some of these big horses, but just as often she rides diminutive Arabs, quarter horses, Morgans, ponies of various descriptions. She rides horses that have been schooled for the jumping arena, ex-racehorses, cutting horses, and once she brought out a horse that had been trained for the sole purpose of dog and pony shows. She believes in the process of dressage, that it will make even the most homely unaccomplished animal more beautiful and capable. In her hands, this is what happens. One of her clients has a small gray Arabian. On a summer weekend Lendon showed the horse in lower-level dressage and won the division. Two weeks later, she reported with great pride, the horse's owner, an endurance rider, won a hundred-mile competitive trail ride with him.

Under Lendon even the most ordinary animal works toward the highest levels, and she is openly sympathetic to the particular difficulties each horse must overcome. For riders, though, her patience is short. It can be stretched only by a student's absolute concentration on the work of getting the lingo down. She is kind and critical by turns as her riders adjust their posture or give up some tension Lendon identifies in a hand or an elbow. She snaps if a corner is ridden lazily in a shallow arc. She suggests, she demands, she implores so that riders working under her can do what she does on the little gray: sit their horses and understand the physical reverberations.

Lendon's gaze falls on another student, another faulty movement. This rider is a young, slight woman on a small stallion. It is a pretty bright bay and delicate enough that I had to look twice to determine its sex. The rider, who is named Karen, is sitting pretty enough, but evidently she is only passively engaged. A few minutes earlier, Karen withdrew a glove plaintively to demonstrate a callus the rein was leaving on the side of her finger. On the finger was a diamond the size of a throat lozenge. Lendon was not interested in the diamond or the callus. She was interested in how the little stallion was going and now she wants more from this rider. She criticizes the way the rider moves the horse through a corner. In a louder voice she demands a better turn. Then, forbidding another turn like the others, she places her own horse in the corner to force the line that will be ridden.

"Karen! How *can* you keep riding this way? I asked you for a *fast* trot. Do you think that's *fast*? Karen! Is he even *trotting*?"

Sweat begins to run down from the corners of Karen's eyes. Her makeup drips to the front of her T-shirt. This is not enough to bring any mercy, because Karen is making no effort to change.

"What are you thinking? I mean, *are* you thinking? Karen . . . *stop* the horse. Get off."

None of the other riders in the arena takes any notice of this. They are initiates, practicing toward the same perfection Lendon has shown them, and each of them has been caught in her sights at one time or another. They know her frustration is nothing personal. It is a question of correct movement. Lendon steps up on the little stallion and puts the question forcefully with her legs against his sides. The stallion lurches forward briefly, then recovers and gives Lendon her answer: intentional motion. He digs in and surges rhythmically along the wall of the arena.

"*This*, Karen, is *fast*." She is talking about speed the way the jazz drummer Tony Williams talks about playing fast, rhythm that responds to a heightened accounting for the passage of time.

"See what I mean?" She dismounts and hands the reins back to her student, who avoids her gaze. "It's very clear, isn't it?"

She gets back aboard Last Scene and rides over to the observation seats and grins at me. "I couldn't let you go home without seeing me really get started." The next time I see Karen she herself will be riding Lendon's Last Scene. I think this is far too generous of Lendon, since Karen is too reticent to ride the horse as he needs to be ridden. But

Lendon says it's not really so generous. Last Scene has a
lot to tell Karen.

Early the first morning I visit Lendon, before the commuter
traffic peaks and Lendon's clients can break free from the
demands of their suburban households, the arena in Bedford is very quiet. I can hear a car go by way out on the
road. I can hear the pigeons that preside from the steel
rafters. Even in her absence, Lendon's voice is a presence.
Her head working student, Liz Britten, is riding, and,
watching her, I see the implications of Lendon's efforts
with horses and students.

   Like other working students, Liz is an apprentice who
assumes responsibility for chores in the stable in return for
Lendon's mentorship. She is very tall, a couple inches over
six feet. Her dark hair is drawn away from her face in a
classic chignon. She is a beautiful girl with wide-spaced
eyes and lifting brows. Liz is much younger than Lendon,
and patience dominates her manner—which is a good
thing, since she spends a large portion of every day riding
young horses, talking pony, repeating herself endlessly. But
before her other riding begins, she rides Medallion, a horse
Lendon has trained to the grand prix movements. There
is some very fancy talk going on, an artful conversation
concerning one of the most difficult movements a horse
can perform, the *piaffe*. Except for the cadenced blows of
the brown gelding's hoofs, the arena is silent. Liz attends

to the horse, a meditative incline to her head, and Medallion attends to Liz with extreme effort. He is teaching her something.

Medallion has performed *piaffe* countless times. He knows how it goes, he knows the steps. But lifting each leg in correct sequence and cadence requires particular balance and intense coordination, and in order to accomplish these movements, Medallion needs to hear the words. They come from Liz's seat and hands, and they remind Medallion of the *piaffe*—ah, *that* move! As soon as he recognizes Liz's intention, he creates the movement. The horse draws himself together, as if he may squat down, but what he does is to remain in place, performing a slow and extravagant trot. It is heart-stopping in its intensity and exhilarating in its power.

Piaffe has been taught since the early Renaissance, but Liz could not know the movement until she had ridden it on a horse that had been schooled in *piaffe*. Medallion has the *concept* of *piaffe*. It is a horse concept, and having learned it from Lendon, he now understands the movement well enough to lead Liz to the physical instructions for it.

Medallion is now absolutely fluent in *piaffe* and the other movements that comprise the grand prix test. Lendon has spent years of physically and mentally demanding work to ratchet up the level of her communication with the horse, and most of what comes out of this work is self-awareness. Medallion is aware of his movement and his ability to assert control over it. When the horse completes a movement and can surge off into the next without correction, he assumes great authority, and when his work is over for the day, he saunters about the arena on a loose rein, his ears forward and his eyes wide. He senses his worth, and he makes it clear that the work is important to him. When it is interrupted, he feels a loss. Earlier that year, Liz had surgery on her leg. She was not able to ride but she continued in her work around the stable. Medallion began to express unhappiness. When anyone would pass his stall, he presented his face in the Dutch door and laid his ears back sourly. This brought no results, so he began hanging over the door to make sure no one would miss his glowering. Then one morning as Liz hobbled down through the shed row, he leaned far out into the aisle, caught her sleeve in his teeth, and pulled her to him. At some level he understands that he and Liz are undertaking something together. He wants to make the moves because movement is the way the horse defines himself. Prowess is his meaning.

It doesn't take much to pervert this meaning. The horse is a big, powerful, and very sensitive animal, and the use of force or even just clumsy expression can violate the horse's understanding. It's just a short trip over the boundary pointed out in a common saying quoted by Danish trainer Bengt Ljundquist: "Where art ends, violence begins," and at virtually every dressage competition where I've seen brilliant riders, I've also seen riders who cause their horses pain and confusion. They lack the physical equipment—they are too tense, they are too fat, or they ride too little—to speak the lingo. They lack the education. They jounce. Their hands jerk involuntarily at the horse's mouth. Their bodies pound the horse's back. Their legs flap at the horse's sides, their spurs digging randomly. They lack syntax. They can't talk pony.

On a cold March day I visited a stable in Connecticut where the footing outdoors was treacherous and forced the riders to crowd into the indoor arena. A woman brought a stiff black horse to the mounting block. The lady was well educated, pleasant, and in fact had a job teaching somewhere. She had just been to a dressage clinic, and the old horse was in for it. He knew it. The white of his eye showed as she climbed aboard. She put her spurs to him and worked on one rein. When he made a tentative effort, she repeated the abuse. In a minute or two the horse was hopping frantically. Even in the frigid arena, sweat broke out on his neck. She worked at him. He threatened to go up. She turned him in a dizzying little circle, a technique she had probably been shown to deal with rearing, but rearing caused by something other than herself. After fifteen minutes the old horse was in a lather and unable to trot forward, walk forward, or move naturally in any direction. The lady dismounted in frustration. "I just don't know what his problem is today." She did not know which questions to ask. She lacked syntax. She lacked even a basic vocabulary. Her old black horse was doomed to meaninglessness.

If you picture this horse and then turn to Last Scene, you will understand why dressage is a tradition of masters and students. It saves horses from bewilderment and makes brilliance possible. Lendon is part of the tradition of mastery. "I didn't choose this, you know," she has told me about riding. "I needed a job." This pragmatic explanation works, but she is too Yankee to explain also that the only reason it works is that, like Keith Taylor, she has the touch.

She grew up in Old Town, Maine, one of four children. Her father built the canoes for which the town is known, and she tells me he was also one of the first people in the country to experiment with water skis. Her mother came from a well-to-do family and brought to Old Town the first pleasure horse the local people had seen. The only people who rode there were kids who piled onto ponies, and they called Mrs. Gray's horse the "piggyback horse." The couple kept horses for the children, and Lendon and her older sister took intense and competitive interest in riding.

When Lendon went to prep school, she found she could make money by teaching people to ride. Later at Sweet Briar, she rode under Paul Cronin, who was interested in "educated" riding and had begun to systematize an approach to training hunters and jumpers based on the so-called forward seat that was becoming popular in this country. It would be several years before Lendon was introduced to the idea of collecting power and balance in the horse's hindquarters. She majored in classics, and after she graduated, she stayed on at Sweet Briar, teaching riding under Cronin and tutoring girls in Latin and Greek. During this period she became interested in event riding, and this led her to Margaret Whitehurst.

An independent and unusual person in her own right, Peggy Whitehurst had been a talented amateur rider with friends who rode on the early U.S. Olympic teams after World War II. She went on to become a commercial pilot for Pan American—evidently gender barriers had not yet been erected in the airline industry—and then married a physicist and settled on a farm in Tuscaloosa, Alabama. Using a couple of Thoroughbred stallions and mares of various breeds, she began to develop a strong strain of homebreds, and by judicious pairing of these animals with top trainers, has produced horses that made the U.S. Equestrian Team lists for show jumping as well as dressage. But when Lendon came to the Whitehurst farm, her assignment was to develop the horses for combined training.

At that time, there was only nascent interest in dressage as a discipline unto itself, but when one of the Whitehurst horses, a mare named Crown Juel, began to make it clear that her talents lay outside running and jumping, Peggy said, "Why don't you try her in dressage?"

Lendon had never seen a horse perform a grand prix

test and still hadn't when she took on a horse named Beppo to try for a berth on the U.S. team that would go to the 1978 World Championships. She shipped the horse to Maine for a few lessons with Michael Poulin, who coached her well enough that she was able to travel with the team to Europe as an alternate. Peggy Whitehurst continued to be a strong presence and support. Lendon took Beppo to the 1980 Alternate Games and around the same time began working with Peggy's Seldom Seen, the first horse Lendon herself trained to the grand prix level and the first of Lendon's "ponies," the small horses she has pitted successfully against the much larger, heavier types associated with dressage.

If you look at photographs of Lendon on horses during this period, you see that remarkably little has changed in the round, generous face, the small chin, the owl rims of her glasses, or the absolute correctness of her position. But details of angle and attitude show her eagerness as well as some uncertainty. How could she have known what she was in for?

At the time, Americans were even less competitive in world-class dressage than they are now. The European riders were backed by centuries of training and theory, bloodlines and study. Dressage as both an art and a competitive sport had been developed extensively—so much so that there were even styles of riding associated with particular countries. American riders had only one Olympic medal to their credit, a bronze won in 1932 by a military officer named Hiram Tuttle. Not many people in the dressage community knew enough to advise Lendon. Michael Poulin, who is known for being brilliant and quixotic, had respect for the classical principles and an intuitive understanding of what motivates a horse. He was based only an hour or so from Old Town, where she had grown up, so she didn't really have to leave home to ally herself. It was physically grueling—she was a working student and rose at three in the morning to be at Poulin's for a lesson at six before she started work—but it was productive. He gave her a lot, and even though she eventually paid off the actual bills, "I never did pay for what he gave me."

Peggy understood the obstacles that faced American riders, and although she was not able to foot the whole bill herself, she helped raise money to send Lendon to Europe to test Seldom Seen in the big time. According to Lendon, the Europeans were completely unaccustomed to seeing a little horse work in the same arena as their own breeds they had developed for the sport. But she says the European judges were more open to Seldom Seen's potential than the judges at home.

Stationed beside one of the first arenas Lendon rode into, a judge—one of three analyzing the ride—watched Seldom Seen's entrance wearily. "He probably thought, 'Oh, these damned Americans—what next?'" He lit up a cigarette and slouched down in his chair to endure the ride. As Seldom Seen was working through the first extended trot, the judge's eyes lifted. He straightened a little more when she rode into the half-passes, and by the time the little horse struck off into *passage*—which is essentially *piaffe* carried into forward movement—the cigarette was out. The skeptic was upright, dictating responsibly to his scribe.

Lendon and Seldom Seen did not take Europe by storm. She was there to watch and learn, to absorb as much of the techniques and theory that could work for the horse and her. She and Seldom Seen acquitted themselves well enough and they returned home to challenge the competition. There was a willfulness in all this, a young American on a pint-sized horse running through donated funds to ride with the elite in Europe. It was a will to know and to master. In dressage, as in art and music and literature —any endeavor where mastery is embodied and transmitted through personality—teacher-student relationships are complex.

Lendon left Poulin's operation in 1987. She was competing successfully with him, but at the same time she realized how dependent she had become on him, that she really missed having him tell her what to do in competition and even in day-to-day training. She says she is determined not to foster that kind of dependency in her own students. She wants them to exercise some initiative and try to work things out with their horses for themselves, as she does with Last Scene.

The little gray is aptly named. He is, in fact, the last of Peggy Whitehurst's competition horses. When I saw him for the first time, he had been in training with Lendon for seven years, and although he was highly educated, he was just beginning to compete in the movements that Medallion knows so well. If he could sustain the physical and mental demands of high-level training, he would develop even further until he too achieved mastery. Along the way, he would compete, and in spite of the hours she has invested in this horse, Lendon's ambitions for Last Scene are tempered by pragmatism. She points out that when it comes to competition, size can be a limitation. A small horse like Last Scene will have to do more—he will have to be more brilliant, and he will have to move with greater amplitude—in order to achieve the same marks as one of the big horses typical of the sport. Lendon met those challenges successfully with Seldom Seen, and it is altogether possible that Last Scene will also be able to make size irrelevant. But Lendon does not let her uncertainty about the horse's prospects in competition distract her from working toward Last Scene's eventual development. This will be a moment-by-moment process, and even as she mounts up, she is making each moment count.

When Lendon swings up on Last Scene and asks him to move off, two things are transformed, the horse and Lendon Gray. Last Scene grows rounder, like a horse in a Renaissance print, more powerful and precise in motion. He gains authority, and as I watch him change, I wonder if Last Scene's experience is something like what happened to me on a dance floor in California. There was a good orchestra, more than forty strings, playing Viennese waltzes. My partner, whom I had met moments before, was an expert on Schubert who had recently returned from a year in Vienna. I knew the box step. He could waltz. He was a strong dancer, and even in the first bars of music, I was waltzing, flying through intricate embellishments of the basic movement as if I knew what I was doing. But I couldn't have begun to replicate the steps after he released me. I don't remember his name or his face, but I remember with precision how his arms felt. His hold gave me authority, and I suspect this is what Lendon gives Last Scene.

Seated on the horse, she is elegant. She has long-legged grace, quiet poise, and when the horse is making his biggest moves, her body barely moves against him. She is carried effortlessly. When Lendon dismounts, she is an ordinary person again. She is no longer statuesque, just a woman of medium height and medium weight. She walks as if she's pushing something—a grocery cart or a stroller —and her hair bobs with every stride. When she looks away from the horses, she talks girl talk. "I *love* the sweater," she tells a client, a jewelry designer, and comments to me about her artistic verve: "Everything she does is that way. She *designed* this watch." She is often impressed by her students' accomplishments in business and art and with the horses, and her students seem to sense her appreciation of them. When they step down from their horses, they appear to leave behind any worries or resentments about the day's work. There is laughter and talk about parties and shows and shopping. The next morning, pony talk will transform them again.

When things work—when the conversation really flows and Last Scene carries Lendon through a pirouette at the canter—whose art is it anyway? *They have no syntax, therefore we may push them around?* Lendon may call the tune, but is she some kind of tough guy waggling a revolver at the feet of a victim—"Now *dance*"? The themes of this literature belong to the horse. They are natural ways of going—trotting, running, acting sexy—made self-conscious. Artifice and the motive to apply it are Lendon's contribution, but no rider could have dreamed up the movement if horses had not first shared their potential to create the movements and to take pleasure from them. A good rider teaches this potential to her horse. A bad rider invites destruction of what comes naturally and could have been heightened by the horse's intentional participation. An indifferent rider creates a hack. A hack is oblivious to art. A hack ignores self-potential for art. What distinguishes Dorothy Parker from Eudora Welty is the difference between a hack and a horse. In much of her work, Parker, amusing as she was, took small things and made nothing of them. Welty saw the potential of small things and made magic of them.

The Riddler, my own sturdy horse with the quizzical blaze, is a hack. It isn't his fault. He is a good enough horse with a good enough heart, but he is ignorant. He talks only pidgin pony, and this is the fault of a series of indifferent riders, the last of them me. My ignorance compounds his. Out on the trail, we move together well. But on these outings we don't talk much. Things are fine until, like the woman trying to ride the old black horse, what I have in mind is beautiful. I ask him to move forward and sideways at the same time, but Riddler's answer is to lift his head and quicken his strides. I ask again. The jaw stiffens, the legs move faster. I squeeze harder with my legs. It is only when he begins to struggle with his head that I finally get the message and relax my grip on the reins. The horse lowers his head and goes ahead, traveling just slightly sideways.

The Riddler may be more prone to misunderstanding than a young horse in a good school. He is older, with memories of humans that cause intense anxiety, and when he is visited by one of these memories, he stops in his tracks and shakes and sweats. Since he doesn't talk good pony, he has only limited forms for expressing these memories and creating happy ones. But a rider like Lendon Gray could expand his expressive range and his trust.

I wish I could—and so do thousands of other people in this country, or at least the thirty thousand members of the U.S. Dressage Federation. It's a pastoral yearning, like the ones that send people out of the city to places populated by other creatures of different minds. People want to howl with the wolf. They want to fly with the condor, swim with the dolphin. They want to ride the horse. But with horses our relationships have become a culture by now, so you can't just buy your ticket and hire a guide. You have to learn the lingo.

Perhaps it is the limitations of so many riders that have brought psychics into vogue with horse owners. For a fee, they will speak the horse's mind to its human. Lendon reported with wonder and wry humor the experience of one of her clients. His game old horse was showing considerable stiffness at the beginning of each ride, but he would go on with his work until the gimpiness disappeared. His owner dialed a psychic in California, and the medium gave a rather lengthy report of what was on the horse's mind. This included two pertinent statements: "I like being part of things even though my feet hurt. I need salt." Almost any aging horse still in work could have made the first statement, but the complaint about the salt applied specifically. When the owner checked the horse's stall, he found that out of thirty-some salt holders in the stable, only his horse's was empty. The experience rattled Lendon enough to make her reluctant to consult a psychic. "I might be horrified by what some of these horses would say about me."

What the horse says *to* her, however, is urgently important. Pony talk may be the language of patience, but it is also the source of Lendon Gray's impatience. She is aware of an imperative need to get the lingo down quickly. A move made wrong is a scale with a wrong note that once played will take hundreds of other performances to make the phrase right and beautiful. She remains keenly aware of the risk of failing with a particular horse, of failing to ride, and this is what makes her raise her voice at a student. "Now. Do it *now. Please.* . . ." The fact that a student is well known or dauntingly wealthy or powerful will not make Lendon back down or even tone down. "I'm getting old . . . *now!*" Dressage is long. Life just may be enough time.

Anne Kursinski and Dynamite (Karl Leck/USESA)

Launching Ferina

A horse flies past my head. The huge animal rides the air, cutting an arc over the single practice rail suspended between the jump standards, oblivious to gravity. When the mare lands, the impact is surprisingly muffled. As astonishing as it is for a horse to launch itself into the air, it seems equally remarkable for an animal of this size to return to earth with so little commotion.

The mare is Ferina. With her rider, Anne Kursinski, she is warming up before a show jumping round in Attitash, New Hampshire. The Fields of Attitash are the level floor of a cul-de-sac valley surrounded by the White Mountains. Bear Peak and Attitash Mountain stand close and huge over the warm-up area, where even though it is only nine in the morning, the sun is hot. That is the way the summer in all of the Northeast has been, unusually warm and bright and dry. As the mare lifts off again, it strikes me that the horse is an improbable partner for a human being, and that this is especially so for the show jumping horse. While Play Me Right is tested by ordeal in the natural world and Last Scene confronts the old and refined culture of the riding school, the world inhabited by the show jumper is a glittering place layered in money and celebrity and artifice. Although recently its glossy surface has been shattered by crime and horse killings, this is the setting that has produced much of the country's most original and brilliant horsemanship. If you look through the floodlit evening interiors, past the candy-striped VIP tents, six-digit purses, and giveaway luxury cars, the goal of all this is really very simple: for a human to fly with a horse.

The show jumping arena is a place as artificial as a theater set. Big fences, each composed of many individual rails, furnish the ring with bright, highly stylized decor. Stripes and alternating rails of contrasting colors are traditional, and these are relieved by theme fences that refer to other times, other kinds of places—a castle, a bank building, an island garden—and sometimes one of these obstacles will incorporate a rectangle of swimming-pool blue water at its base. The uprights that hold the rails are decked with shrubbery and flower arrangements, and between classes when the fences are moved, the young men on the jump crew run back and forth with the heavy flowerpots. Steve Stephens, a Floridian with down-home humor and an earlier career as a successful grand prix rider, is responsible for designing the course Ferina will jump, in fact all the courses that go up during the two weeks of Attitash. His bright, imaginative decoration of the course camouflages the stiff athletic test that the horses confront. Stephens has considered what will make an appropriate challenge for horses with this level of experience. To test the speed, strength, and maneuverability of the horses, he has set every fence at a calculated angle and height, at a precise distance from the fence ahead of it.

The object is to ride over all the obstacles as fast as you can without bringing down any piece of a fence. If a rail falls, you are penalized. If your horse is too careful and too slow, the clock punishes you. The bigger the fences and the more intricately they are spaced, the bigger the purse. The size of the arena and the relative scale of the horse make this game roughly analogous to racing a car on your veranda.

Ferina, or any horse, is an unlikely package for this project. Her body is an enormous barrel that weighs more than a thousand pounds, and inside the barrel there is always potential trouble. Ferina lives in bondage to her digestive tract. It never stops working, and to avoid a life-threatening bellyache, the mare must send a steady supply of roughage through her gigantic tortuous gut. This is a very large, very dense, and very delicate system. It is propped up on four narrow legs, where Ferina's bone and muscles and tendons are protected by only a thin layer of skin.

The arrangement is top-heavy, and because about 60 percent of a horse's weight is in its front end, it requires tremendous thrust from behind to lift it off the ground. But the disadvantages in the engineering of the horse don't trouble riders like Anne, whose everyday work is to launch horses over obstacles as tall as, sometimes taller than a human. Anne is an experienced pilot who has spent enough time in the air over fences to have an accurate inventory of the horse's physical potential, and specifically, she has taken account of the abilities of Ferina, whose jumping competition this summer is one phase of a carefully plotted career trajectory. Ferina is a German-bred mare Anne brought to this country to develop for the Olympics. This summer "the Road to Atlanta" is becoming a catchphrase in the equestrian press, and while the next Games are much on the minds of people in all three Olympic horse sports, interest and speculation seem most intense among the people who follow show jumping.

Ferina's warm-up, first simple figures on the ground and now over fences, is to let her limber up, to try out the muscles and perception she will use in jumping, and to prepare her mentally for the jumps that will confront her in the ring. Anne has a strategy for this process, and after the practice rails go satisfactorily that day, she lets Ferina meander on a long rein, head down, toward the in-gate. This brief passage leads to fences the mare will have to jump more than satisfactorily. Even though it is a class without a big purse, Ferina will have to jump them fast.

Because her round is early on a weekday, there are not many spectators to watch Ferina or her competitors. Anne tells me that, in general, horse show audiences in this country do not approach the crowds at the major shows in Europe, where show jumping has a much longer history and a following second only to soccer. In this country, according to her, only the jumping at the Devon Horse Show and the American Gold Cup draws the same kind of crowd and enthusiasm. This year Attitash is better attended than the show in Lake Placid just a month earlier, another big-time jumping competition in a gorgeous mountain resort. At Lake Placid, the sunshine was stifling, the heat and dust relentless, and while nine hundred horses competed there, only a few people came to watch. At Attitash, the heat remains intense, and during the week the VIP tent is where most of the spectators and competitors gather. On the weekend, on grand prix day, some spectators will file into the metal grandstand on the other side of the arena.

By eleven or so the tables under the blue-and-white VIP tent that extends the length of the arena will begin to fill. This is adjoined by a small tent where you can buy food—salads, sandwiches, meat from the grill—or drinks from the bar. At horse shows, people seem always to have a paper plate or plastic cup in hand. For those who don't ride, the summer is an extended house party. For amateurs who do ride, it is rounds of competition interspersed with rounds of spectatorship and table hopping, and for professionals, the VIP tent is a place to meet clients. The people here are more obviously well-to-do than the three-day crowd. Their style begins with prep school—linen blazers, ponytails, baseball caps, souvenir T-shirts from Curaçao, sneakers, sandals, espadrilles, and other shoes that would be ruined by a trip to the stable—and their mood is genial. Compared to fans of eventing or dressage, this is a carefree, fun-loving crowd. They bring the kids. They are not so strict with themselves about personal fitness—there are many smokers here—or serious about practical gear. I have occasionally seen amateur show riders exercising their horses in shorts and sneakers. Because competition is almost incessant, they learn how to find down time and take their fun. They travel the same routes, eat in the same restaurants, share the latest jokes, and follow the performances of an elite group of horses over the long season.

They know Anne is one to watch and they will keep an eye on the big clock as Ferina jumps off over the shorter course. At the in-gate, competitors and grooms come and go to watch the rounds, check out the strategies of other riders, note the clock. Just beyond, seated at a desk under an awning, there is a second announcer, who, while the official show announcer introduces riders to the spectators and pronounces on their scores, broadcasts to the warm-up area, giving riders the order of "go." Although Anne has likely seen a schedule for the rounds, there is a great deal of flux in this list. It changes constantly, and it is the caller under his awning who gives her the authority to appear at the in-gate.

If you sit in the stands at the shows only occasionally or watch Ferina's round on television, you might not see at first how intense this exercise is for the horse. While show jumping is the most readily understood and easily televised of the Olympic horse sports, its confinement to a relatively small area belies its speed and risk. The horses and riders competing at this level reinforce the deception. Anne and the other top U.S. riders who have come to Attitash—Michael Matz, Norman Dello Joio, Margie Goldstein, Chris Kappler, and Debbie Dolan—don't often make mistakes, and their horses don't often dig in at a fence or try to dodge out. But the sport calls for the same kind of nerve and precision as downhill skiing. Every course demands a great deal from the horse as an athlete, and one wrong move can produce a truly scary crash. A lot goes on in a period of less than a minute and a half, and it's easy to miss the action unless you can see it as the horse and rider would, fence by fence.

Aside from the flowers and shrubbery around the fences, nothing that faces the horse is natural. At the in-gate at Attitash, Ferina has no specific knowledge of the test ahead of her. She is just eight and has not yet had grand prix experience, but in all probability she is aware some kind of challenge is imminent. Anne, on the other hand, has already seen the fences and turns close up. As soon as the course designer approved the height and spacing of the obstacles and such internal details as the distances between rails, the course was final, and the competitors toured it on foot, pacing off the distances between fences, weighing alternative routes between obstacles, gauging the angles for their turns. Anne strode back and forth between some of the fences more than once. She is a lithe blond woman in black boots that come to her knee, white breeches, and a blue shirt with a choker collar. She has a quiet intensity. Walking a course or standing beside a schooling fence with her hand on her hip, she is completely engaged. Even when her body is still, she gives the impression of activity.

When it is Ferina's turn to jump, Anne moves her into the ring in a carefully rated canter, a slow, emphatic run-up through the fences she wants Ferina to see for herself before they start. She is poised over the horse, half out of the saddle, and each time she enters the ring, she will ride this way, curled slightly forward to monitor the horse's stride. The posture and rhythm are characteristic because this procedure is systematic, like nearly every aspect of her dealings with horses. If you are searching a crowded warm-up ring for Anne, you can identify her and her horse by their silhouette.

Under saddle, the statuesque Ferina fits into the Kursinski outline, and you do not notice so much how beautiful the mare is. In spite of what I have said about horses being top-heavy and overloaded on the front end, Ferina is not at all ungainly. In fact, she is exquisite. The shape and proportion of her quarters, the width of the bones and the angles of their joining in her legs, the muscling along her back, the lift and arc of her neck, the refinements of her leaf-shaped ears and large luminous eyes—they all harmonize into a horse that the English painter George Stubbs would have placed at the center of one of his animal arrangements. Although Stubbs adhered assiduously to the anatomical realities of the horse and even published a treatise on the subject, many of the images that resulted were to some extent idealized representations of a mythic notion of the horse's body. It's just not that easy to find physical perfection in a creature that really eats. When I mentioned Ferina's beauty to her groom, Stacy Falco shook her head, apparently still wondering over it herself. "I know. She's *perfect*."

But to earn her keep, Ferina has to do more than look good. In the first round, she has to be confident and careful and efficient. If she makes the required time and leaves the fences still standing, she will jump a briefer course at the fastest pace she can manage without leaving any rails on the ground.

There are plenty of rails in the first fence. They are striped in bands of red and white and hung one on top of the other. As is the custom for first fences, this one is straightforward. There is nothing tricky or technical about it, and Ferina lifts off obligingly. With her legs, she does exactly what is wanted. She pulls them up, the way a plane retracts its landing gear, front legs slightly ahead of her, hind legs slightly behind, and lower legs tucked snugly against the upper. Even before the mare lands, Anne is looking for the next fence, which is placed on a very mild curve from the first obstacle. In the opposite corners of the arena are large digital chronometers, and as Ferina gallops and jumps and gallops, the seconds blink past.

The face of the next fence is solid and dark, and the fence has a back as well as a front. The two elements make the obstacle wide. Ferina will have to launch out more, stay in the air longer before descending. At twelve hundred pounds, she outweighs Anne by at least ten times. But Anne's human body is handier than Ferina's. As many technical works on riding point out, the only top-heavy element in the human body is the head, which physically, as well as neurologically and intellectually, is critical to riding. Otherwise, the human body is slim and long-legged, quickly flexed and prehensile. Anne can sit and bend, lean and reach, wrap herself around the horse and operate on each side of the horse to regulate Ferina's speed, guide her to the right takeoff point, and regulate the force of her launch. The scope and force of Anne's movements are minuscule compared to the ones she asks of Ferina, and this is to me one of the continuing mysteries about riding horses, how little effort it can take.

The mare sails across the dark wide fence, comes down galloping and turning toward a fence made of narrow planks, each painted to represent a section of the American flag. Because the flag is supposed to be waving in the wind, the planks have been sawn with curves in their contour.

The mare wants to pick up speed. But this is not part of Anne's flight plan, which has already been filed for the American flag. It is filed in the passage of a second. The progression of these instructions, no matter how instantaneous, is methodical, something Anne gave thought to before she made it routine.

They are at the Western town fence, a solid-looking three-sided corral. Fences like this one are there partly to delight the spectators and partly to distract the horse, because Ferina does not see her surroundings the way we see them. What she sees best are objects that are far ahead of her, and just as we do, she uses binocular vision to unify the images the two eyes hold of the same object. But each of Ferina's eyes can also independently see objects on either side of her. She cannot coordinate the peripheral images received by each eye, and in that sense she sees double. She sees the rails ahead of her at the outlet of the corral, and at the same time she receives images of the walls of the obstacles on either side of her. Blinders were invented to block off just this kind of distraction for carriage horses.

The rustic colors of the Western town fence present another kind of hazard for Ferina. They are primarily wood tones, taupe and brown and gray, and while Ferina receives some color in the images of the fence, it is much less distinct, more grayed-out than the color we see. Steve Stephens compared her perception of color to our perception of the picture on a black-and-white television. This means the Western town fence probably doesn't stand out much, one way or the other, and its coloration could cause Ferina to misjudge its size. But sixteen or seventeen seconds into the round, she lands safely outside the Western town.

The line takes a wicked turn to confront a very tall green-and-white fence. Even if Anne guides conservatively to the most ample version of the turn, Ferina will still be taking off just as she is completing her turn. Her steering must be more responsive than the most refined automotive steering. Ferina actually cuts the turn and takes off. In the air her rider is planning another turn, the one that happens just six or seven strides after the horse lands.

If you could look under the helmet into Anne's face now, you could see the changes that begin to twist into her expression as soon as she rides into a ring. One side of her face screws down, squinting, drawing back, while the other opens and the brow lifts. Although it looks like a wicked-stepmother expression, it is brought on by intense concentration. The turn six strides out from the big green-and-white fence happens because Anne has planned for it to happen, and she has a method for enabling Ferina to nearly double back and be looking straight through a three-part fence that's less than a length away. Having observed that steering is one of the abilities it takes to win, Anne has incorporated it into a program to build Ferina's skill.

Having observed all the essentials of winning, she works through rational process for what she wants.

The fences are coming up faster. It is the triple now, three red-and-white fences separated by distances a horse cannot automatically take in stride, and the third fence is actually two fences set together with a little spread between them. Ferina seems to scramble a bit just before this last, wider element, but she takes her rider up and out. They are turning again.

Although this course is something less than the highest level of difficulty, it still makes heavy demands on the horse. These fences, at these heights, set so close together, are no project for Keith Taylor's Faktor, who has not yet put to simultaneous use all the skills necessary to navigate them. Faktor could jump any single fence in the arena and perhaps any two. But at his stage of development, the successive demands of the thirteen obstacles would be overwhelming. Even a horse as bold and experienced as Play Me Right would have to do some serious homework to attack these fences in this layout and come away without being humiliated.

Oddly enough, of the horses I know who are not show jumpers, it is Last Scene, the small gray horse Lendon rides, who comes closest to having the skills to succeed at a task like this. He has the compression and the maneuverability. He has no actual jumping experience, but he can draw back on his haunches and turn at the same time. He can adjust the length of his stride incrementally and instantaneously switch the side of his body over which he is balanced. He can surge forward with nearly any degree of force.

There is a great deal of dressage in the turns, shifts of balance, and changes in impulsion that Ferina and other show jumpers make. It is not so refined or precise or technically difficult as the movements that Last Scene works on. It is broader and faster, and it is called flatwork. I was aware that many successful show jump riders put a lot of effort into this, and I think Anne is especially attentive to the details of dressage—in her book *Anne Kursinski's Riding and Jumping Clinic*, she devotes as much advice to these aspects of riding as to the processes involved in actually getting the horse over a fence. But I was surprised by the degree of emphasis all show jump riders give to "flatting." I visited one lovely hilltop farm in upstate New York that starts and sells many world-class jumping horses. The stalls were finished in natural wood and afforded the horses views to both the inside of the stables and the activities on the hill. The tack room and its appointments were gleaming, and the horses were vacuumed the way more ordinary animals are brushed. Every detail of the operation had been thought out, but nothing about the place revealed that it produced jumping horses. The only "jumps" in evidence were a few poles laid on the ground in one of the riding areas.

"A jump is just a stride" is one of the truisms that

accounts for controlled amounts of drilling over jumps. "A horse only has so many jumps in him" is the other. Jumping puts a horse's legs and back under increased strain. In any horse sport, legs seem to give out more frequently than any other part of the horse, or at least to be named as the source of pain most often. In spite of their power, a horse's legs are thin, rather delicate supports for a massive body. The skin and hair covering the bone don't offer much protection from the bangs and dings a horse takes when he jumps or when he plays at pasture. They don't even protect the horse from his own clumsiness. Mostly, though, the serious problems, the ones that remove a horse from contention, are caused by concussion.

When the horse gallops and jumps, a lot of weight and force are sent down through the narrow shaft of each leg. The hoof is the primary absorber of this shock. It is a tough case of material like fingernail or turtle shell that encloses the foot bones and soft tissue around them. The rim of the hoof tends to be shaley, and while it has evolved to wear well and replace itself on the horse traveling at the distances and speeds that occur in nature, the hoof wall tends to chip and flake and crack with increased impact and use—hence, the need for horseshoes and, often, for pads that cover the bottom of the foot. The hoofs and a good farrier can do a lot to minimize the pounding a horse's legs take when the animals runs and jumps, but repeated concussion still takes its toll. This is why it's not unusual for a horse like Play Me Right, who has been galloping and jumping for more years than many of the horses who run against him, to have chronic leg problems.

When Play Me's soundness seems questionable, Keith takes the horse to swim in a pool built in West Chester to exercise racehorses with injuries. When you look down over Play Me as he swims in the moat around a concrete island, his body floats like the body of an insect over the steadily churning threadlike legs. "Skinny legs," the man who owned the horseshoe-shaped pool told me. "That's what makes them have to work so hard." Play Me blows noisily, getting a workout without coming down on his legs. His nostrils are huge. You can see deep into the flange, beyond pink to dark rose. His breathing is full and strong and rhythmic but the cadence is oddly rapid, unsynchronized with the movement of his legs.

Galloping in the ring shadowed by the mountains at Attitash, Ferina takes a breath with every stride. She does not breathe during the effort of jumping and this taking of breath, checking of breath, is syncopated with the blows of her feet as she takes the course. Ferina does not need to maintain the intense level of cardiovascular fitness that Play Me and Faktor condition for because she competes by running and jumping for only a few minutes. But she is conditioned for strength and suppleness and speed in tight places, and she is fit.

Between the last two fences is another switchback,

somewhat more elongated. It runs from a blue fence, three gradations of blue—aqua to delphinium—to meet an innocuous-looking but very wide and very blandly painted set of yellow and white rails. Evidently, in spite of the deceptive color scheme, Ferina sees this fence for what it is, and the chestnut legs retract and snap into a tuck as if they were spring-loaded. When the mare touches down over the last fence and pounds past the electronic eye at the finish, seventy-eight seconds have accumulated. Ferina has gone clear within the time allowed, and she remains in the ring to take her jump-off round.

Ferina is just the first of Anne's rides today, and while Ferina is approaching the start flags for the jump-off course, Anne's barn manager, Carol Hoffman—Hoffy, no one calls her Carol—is leading another horse up from the stables to the warm-up, a bay named Top Seed. Hoffy is quite tall and thin, angular. Her blond bob is usually punched back under a baseball cap, and she is usually smiling a big, generous smile. All the time Anne is warming up horses and competing, work continues at the stables, where Hoffy is a bright force behind a tight organization. Although she is an accomplished rider herself and shows her own horse, her primary job is to keep Anne's Market Street Stables running efficiently. She attends to the details of the books and billing and ordering, and she works with the grooms to be sure the right horse appears at the right time, wearing the right equipment, having eaten when and what has been specified.

Like the jumping course, the spectator tents, and the boutiques and concessions, the stables behind this set are also a temporary construction. They are portable, canvas-sided stalls set under a circus tent. In contrast to the festive formality of the jumping course, this backstage area is something like a medieval barnyard. There are mountains of soiled shavings between the tents, a steady traffic of grooms with buckets or wheelbarrows or horses, dogs of every description engaged in a subcurrent of dog errands, and puddles left after horses are hosed down. Professional grooms, who could find positions only at the very top stables in eventing and dressage, are regulars at all levels of show jumping. Some are hopefuls, riders working their way to the big time. Some came to be grooms when, young and without much education, they found something they were good at, and some came because you can do this work without having to speak or read much English. There are many newly arrived Latinos on the circuit, Puerto Ricans, Mexicans, Guatemalans, Costa Ricans. Grooms are a separate class. They wear T-shirts and jeans with a rag hanging out the back pocket. They carry buckets and fly blankets, and during the shows they come to work at 4 a.m.

After Ferina's jumping round, the beautiful mare will be bathed and wiped down. Her legs will be iced, then poulticed to tighten up the tissues. Later, she will be walked to stretch her muscles. She will be fed. Her water will be replenished, and the shavings in her stall will be picked out. Manure is removed immediately. Ferina is immaculate. In the mornings, she and the other horses breathe the mist from a nebulizer to clear their lungs and throat of any dust from the riding areas or bedding, any pollen that might float in the air at the place the horses are showing this week.

Ferina's care is constant, professional, and expensive. You cannot consider this sport without considering money. A top horse can cost a million dollars, and the bills for any horse that competes regularly are heavy—the figure I've seen most often is fifty thousand dollars a year. But the people who pay the bills are not usually the riders. The money and the athletic ability come in separate packages. Frank Chapot, six-time Olympic veteran and now coach of the U.S. team, pointed out that the general public has the misconception that show jumping is a sport in which only the elite can participate. While the horses are owned by the very wealthy, the majority of the top riders come from middle-class homes, and even at the peak of their careers, their earnings are less than if they worked up to the top ranks in tennis or golf.

Travel is a big item on any show jumper's bill. It is an essential condition of the sport. When the last round has been jumped in Attitash, the jumps will be disassembled and the rails and planks and pieces of scenery will be loaded on a tractor-trailer like so many Tinkertoys and shipped to the Hamptons on Long Island, where there is a show the following week. Another set of trucks will haul in the portable stalls and the tents to shelter them, and the last wave of vehicles, some semis, some heavy pickups with trailers, will bring the horses. Each show is five or six days long, and its date and location are fixed in the calendar year. All summer and fall the action moves up and down the East Coast, sometimes taking a bounce as far west as Detroit or Nashville, and on the West Coast the action is roughly parallel. When the weather turns colder and wetter, the competition moves indoors for shows like New York City's National, Harrisburg, the Washington International, and the Royal Winter Fair in Toronto. Then, after a break for the holidays, the jump crews start setting up in Florida for the winter season.

The shows move, and the horses move with them. They are trained and conditioned in the same places they are shown. Any horse that competes must travel. For Last Scene and Faktor, being loaded into a trailer is a common interruption of their routines at home. For Ferina and her stablemates, travel is their routine. This includes air travel. Lendon and I talked once about what this experience was like for the horses, and she described how they are loaded into a shipping container much like the interior of a trailer. This is rolled out on the tarmac, raised up to the body of the plane, and pushed in. She thought the moments when the crate was being elevated might be worrisome to the horse, but as for the thrust and noise of takeoff and the sensations of landing, she said, "I can't imagine it's any worse than being in a trailer bumping along the Long Island Expressway."

There are horses who ride around in trucks and trailers as nonchalantly as many people take the bus to work, and there are horses who don't "travel well," who are made nervous by the ride. Sometimes this is reason to ship a horse early to a competition and give it time to readjust before competing. Interestingly enough, sleep, the function travel most often disrupts for humans, may not be changed very much for horses, which have unusual sleep habits.

Ferina's ancestors were preyed upon, and I suspect this accounts for the insomnia she shares with other horses. She gets most of her rest while standing upright. She is hypersensitive to sound, and much of the time she isn't called upon to be active, she spends dozing, locking one hind leg to support herself, maintaining just enough awareness to be aroused by anything that signals danger. The only real sleep she gets is a half hour or so found when she actually lies down, and although horses usually take this deep sleep sometime between midnight and dawn, it is not unusual to find show horses lying asleep in the middle of the afternoon.

Accustomed to life on the circuit, Ferina seems inured to continual change in the world around her. She has all her stablemates with her, and she is familiar with many of the horses that ship into any given show. She recognizes the geometry of the show's layout, the warm-up area, the in-gate, and she understands the ring itself with its obstacles constantly shuffled into new arrangements. Part of Ferina's education takes place here in New Hampshire, just as it does at Lake Placid, the Hamptons, Devon, and the mare is just as much at home in Attitash as she is in Pittstown, New Jersey, where Anne's Market Street Stables are officially based.

Market Street is named to honor Anne's mother, who had grown up in York, Pennsylvania, the daughter of a "big-time" horse and mule dealer there. Market Street ran through the commercial district. It was the road to get on to go someplace bright, and although they settled in Pasadena, California, Anne's family maintained close connections with the East Coast. Anne's mother showed horses as an amateur and started her daughter in riding lessons at a stable run by Jimmy Williams when Anne was just four. Williams was an innovative horseman who started his career with Western horses and trained horses for stunt work and the movies. Then he turned his attention to the tricks of getting a horse to jump big fences fast and clean. He was very successful in training for this task, and by the time Anne was put up on one of his horses, his cow ponies were gone. Her first lesson was the beginning of a long apprenticeship.

Except for a period during which she studied with Hilda Gurney, the best-known practitioner of dressage on the West Coast and a contemporary of Michael Poulin's, Anne worked with Williams until she came East. Gurney encouraged Anne to think that she could become competitive in the world class, that she rode well enough to acquit herself against the Europeans, and Anne stayed with Gurney long enough to bring one horse to upper-level competition. Then it was back to Williams, and before she reached twenty, she had become one of the most successful show jump riders on the West Coast. Still, she wanted more. In 1978 she rode jumpers at Spruce Meadows in Calgary and caught her first glimpse of competition at the international level. West Coast riders had rarely reached this level. According to Anne, Mary Mairs Chapot, another Williams student, was the only West Coast rider who had won a spot on an Olympic squad. But the idea was there, and it had been for a long time.

"Even as a child, I always had this little dream"—as if she were grasping something very small with the tips of her fingers, she showed me the location of this seed of ambition, somewhere near the solar plexus. "I didn't talk about it, but it was always there." When she told me this, I thought of the girl in *National Velvet* who keeps a box with tiny paper horses in it, and the night before she leaves home to race in the Grand National, Velvet brings out the box to have a last look at the fantasy horses inside.

George Morris, a New Jersey rider and coach with substantial international experience, encouraged Anne to think of moving East. Morris was and continues to be extremely influential in this country, not only because he has been widely published but because he has brought along a legion of professionals who have been successful in diverse aspects of the sport. He just happened to have a horse he wanted her to ride.

Moving East was a big step for Anne to take. It meant not only distancing herself from her California teacher and friends but establishing herself commercially in show jumping. Behind every show jumping rider there is an enterprise that sustains the rider's competitive life, and when Anne left the domain of Jimmy Williams's cowboy savvy, she left a job to start a business. She needed clients with financial depth, she needed top horses, and she needed real estate and a support staff. Although Lendon and Keith—all professional riders—have to make a business of their riding and teaching, the life of a successful show jumper has an executive component.

Gathering sponsors and horses that can win for them is an iffy business, but aside from independent wealth, it remains the primary means of access to top horses. "It's kind of crazy," Anne pointed out. "I call up the lady and ask if she will spend thousands of dollars so I can ride a lovely horse." But this has been necessary, and it has brought a succession of what Anne calls "lovely" horses. She is not referring to their beauty but to the way they jump and turn and gallop, the way they handle, and how careful they are. By 1984, Anne had Livius, who carried her to an alternate's spot on the team for the Los Angeles Games. By 1988, there was

Starman. At the Games that year, Lendon Gray rode on the team representing the United States in dressage, and Anne rode Starman for the show jumping team. The team came home from Seoul with a silver medal. Then in 1992, she went to Barcelona with Cannonball, a very good horse but not yet seasoned enough for the challenges of Olympic-level courses. Anne left Barcelona disappointed, but characteristically, on the plane returning across the Atlantic, she quickly devised a plan to go to the next Games with more horse power.

Managing Market Street, raising funds, and looking after clients make heavy demands, and fortunately Hoffy is organized to meet many of these. Still, Anne cannot ride without at the same time devoting a good deal of energy to the institution that keeps her in horses. While many show jumpers are owned by individuals, the costs of ownership have become so high that these horses are increasingly syndicated the way racehorses are. After Barcelona, Anne's plan was to avoid the legal complexities of limited partnerships by asking her best clients to form a simple group to buy a horse for the Olympics and share expenses. Trust among the owners would be essential. She wrote the clients to ask if they would be willing to be part of the team. "What other kind of athlete can you think of," she asked me, "who has to put together something like this?" But she is genuinely appreciative of her clients, and I suspect that her candor and her uncomplicated drive to win —not just a spot on the team but the Games themselves, and not just win but win and at the same time be all that is good in a winner—are convincing to the people with the money.

It's a fragile enterprise. Funding can get shaky. A horse can go lame or go off mentally. Anne can crash and fall. She can be injured and put out of action, as she was early in 1995. She lost a stirrup during a very fast turn, spun off the horse, and landed so hard that the resulting muscle spasms kept her from sitting on a horse for a couple of weeks, a period when all the top riders were earning points to be named to train with the U.S. Equestrian Team in Europe. The USET is the organization responsible for fielding teams for international competition in all three Olympic sports. There had been a great deal of discussion about how essential European experience would be to success at the Atlanta Games, and when Anne was kept out of competition during this crucial period, she began to make plans to ride abroad independently. As it turned out, however, late in the spring she won the grand prix at the Devon Horse Show, which gave her enough points to make the USET list.

Ferina is one of three chestnut horses Anne is developing for the Olympics. Her male stablemates, Eros and Dynamite, have both been successful over grand prix courses. Dynamite won the grand prix at Devon, and at the same show grounds Eros captured the 1994 American Gold Cup.

Ferina is in New Hampshire this weekend, not Rotterdam, because two days before the U.S. Equestrian Team trunks were packed and put on the van going to Kennedy Airport the mare twisted a front leg trying to right herself from an awkward sleeping position. She is the most junior member of the squad, the least experienced, and she will have to progress fairly rapidly through more difficult courses if she is to become truly competitive for the Olympics. But expediting training to meet an end is not part of what Anne calls her "program," so she decided there was not much to lose by keeping Ferina home and letting her heal. Dynamite and Eros followed their groom Rob Spatz onto the van, and Ferina stayed behind at Lake Placid with a clean white bandage on her tender leg. Anne has just arrived in New Hampshire from riding with the USET in La Baule, and before the Attitash show climaxes with a grand prix, she will be back on an airplane bound for Rotterdam, where Rob will meet her with Eros and Dynamite. Late in August, they will all return for the shows at the Hamptons and Spruce Meadows in Calgary. Then in September, Eros and Dynamite will get on the Market Street van and ride just a couple of hours to Devon for the American Gold Cup.

It is a summer of Atlantic crossings and a long season of shows, intense heat, and drought at home. Anne works the same long, physically demanding days, the same endless workweeks as Lendon and Keith, or any professional. Their weekends are committed. They do not take days off. They don't even take lunch. International travel multiplies these demands. Personal life must be put aside. "You can be a professional and have a marriage, a family," Lendon observed, "but it's very difficult to do that and be an international competitor. The other person has to give up so much." For Anne, tight-knit relationships with Hoffy and the Market Street staff provide a familial support during the summer's single-minded campaign. Anne needs to do well with the horses on both continents and to continue doing well through the winter season in Florida and the spring outdoor shows on the East Coast. Each rider who competes for the United States will be selected for the team in combination with a particular horse. She wants to qualify with at least one of the horses for a place on the team and, beyond that, a place on a team that can win, the way the Americans won in Los Angeles.

Since their gold medal in 1984, the Americans have not been a strong force in show jumping. The streams of money that riders were relying on for sponsorship began to be diverted by interests that worked against athletic achievement. The people with the money, the people who bought the horses and paid their bills, developed a strong desire to compete themselves, and there were professionals interested in doing business with these wanna-be grand prix riders. At the same time, there was contention about how the U.S. teams were selected, and one of the turning points was a lawsuit against the U.S. Equestrian Team by the rider Debbie Dolan claiming damages from the selection process that favored Anne Kursinski for a berth on the 1990 world championship team. Partly as a result of the suit, the USET's selection process for show jumping is now completely "objective"—that is, based on points and not the opinions of committee members about the capacities of a horse and rider—and it relies on a long series of trials held in conjunction with grand prix competitions. The litigation and a subsequent challenge for control of the USET were symptomatic of deep divisions within the show jumping community, and morale was eroding. The lowest point for the sport came four years later when a sweeping series of federal indictments charged more than twenty people with business interests in the sport with arranging the killing of horses in order to collect from their insurance companies. Within the horse world and the equestrian press, there was uproar. There was indignation, concern about fallout from sponsors, and great sensitivity about how any "horse person" would be regarded by the general public. "If it happened," Anne told me, "*say* it happened . . . and let's go ahead." She is not one for complicated fixes, and she is pragmatic about the rigors of the trials facing all the riders. "We need to pull together if we're going to put together a team that can really do the job."

The job is winning. Winning is the way you get the points that put you on the lists that determine standings and selection. Winning is the way to move up the lists, survive the last cuts. Because the rider must qualify in partnership with a particular horse, Anne needs to keep each potential qualifying mount—Dynamite and Eros and Ferina—as close to the top as possible. She needs to win, and as the daughter of a schoolteacher with strong ties to the church, Anne seems to assume that there are ethical components to the task—hard work, conscientious horsemanship, aboveboard dealing, consideration of others, and a life that withstands scrutiny. This is the way she answers my questions. It's a fairly wide-eyed philosophy for a world where money has bought so much and where cynicism threatens to take over, but it seems to work for her. When Anne tells people she wants to win, they want to help her.

The top show jumping riders are so slick and their horses so canny and competitive that any of their rounds can lull you with the impression of something practiced often, well rehearsed. But every time Anne rides Ferina—or Eros or Dynamite—into the ring, she asks them to try something different, something they have never done before. Steve Stephens's course at Attitash is not the same as the one he will set out at the Hamptons or the one the Californian Robert Ridland will conceive for the American Gold Cup. Lendon Gray knows the movements and sequence for each test she rides with Last Scene, and theoretically, she could practice the test over and over. While she won't drill the test in its entirety, because Last Scene would learn the test and begin to anticipate the movements, she can rehearse moves, sequences, and transitions. To some extent, the event horse can preview the challenges of the cross-country course. Often the horse is allowed to "school" a particular course. This trial run usually takes place some weeks in advance of competition over the course. It is not for the purpose of memorizing the trip over the obstacles but to allow the horse experience with any special problems that particular course might present. But the fences are permanent. They stay where they are, asking the same questions of the horse.

Each show jumping course asks something slightly different of the horse, and what the show jumper can prepare for these challenges are skills and strength and nerve. Each show round provides a different experience, and in that sense Ferina's jumping round at Attitash is as much training as it is competition. To save wear and tear on the horse's body, Anne doesn't often jump Ferina as high in training sessions as she jumps in the show ring. She reserves the mare's biggest efforts for competition, and she assesses the course at a particular show, anticipates the propensities of the course designer, and adjusts the entry for Ferina accordingly. She wants each experience in competition to demonstrate to Ferina what she can do.

The jump-off will be a speed round over only seven fences. Anne gathers Ferina back into the characteristic profile and rhythm with which she approaches a course. As she gallops toward the markers for the start, there is more surge in the stride, more drive from behind. Each stride is a leap. There are only seven fences ahead of her, and now the only time limit is the one determined by the fastest horse. With the mare bouncing under her, Anne flattens a little along her neck, and they slice any excess from the turns. The seconds flicker past, but the mare is moving quickly too. The round is tight and fast. Then, at the American flag fence, which comes next to last, the top contoured rail is struck by Ferina's hind hoof and falls. She flies out over the last fence and drives across the finish line. She has lost too much with that rail.

But this is just one round. Once it is over, Anne's face relaxes, and even though the helmet still hides a lot, you can see that Anne, like Ferina, has physical beauty. She has pale blond hair and large violet-blue eyes set in a face whose contours could be photographed to sell cosmetics. But she does not seem to be aware that she is beautiful. Her expression at rest is open, unselfconscious, and candid. It is likely that already she is looking back on the round, assessing its implications for the horse and for her own goals.

Anne's focus on the Olympics makes her work with horses qualitatively different from Keith's and Lendon's. There is so much pressure: pressure for the horses to compete successfully, pressure to prepare them for the very big courses and very talented horses with which the Europeans have set the standard, and at the same time, pressure to protect the horses from crumbling or peaking too early under too much pressure. But she deals with this by making it a conscious part of her strategy, because managing pressure is part of winning. She has won by figuring out how to win, then teaching herself to do that. I have never met anyone who wants to win as much as she does, but there is something remarkably open in her attitude about the process. In spite of the fact that show jumping is a forbiddingly expensive sport, in spite of the fact that the investment must go on year after year while you learn to ride a little faster, a little cleaner, a little smarter, she lives by the belief that winning is something you get by working at it. There may be an elite of riders with innate talent, but Anne doesn't appear to recognize this phenomenon. Do it. Analyze it. Do it again.

She gives Ferina a consolatory pat, and they take the gate to head back toward the warm-up area. "Ask for a lot, be happy with a little" is something Anne likes to say.

Meeting Their Minds

When Faktor reared and bumped heads with Keith Taylor, he broke Keith's nose and sent him into the hospital for three days. The bill came to twenty thousand dollars. Although insurance paid for three-quarters of this, the remainder was overwhelming, and worse yet, Keith was prohibited from riding for a month. But he accepted this restriction for only two weeks, and since he spent most of that time running alongside his horses to work them in hand, grounding Keith had little effect on his recovery. I ask him how he is managing the hospital bill. He looks down from Faktor's back with a big smile. "Twenty dollars a month—and every month they send me this very nice letter: 'Thank you, Mr. Taylor.' "

"Do you hold it against Faktor? I mean, he hurt you, put you in debt."

"No, he didn't know what he was doing."

The question of what a horse knows and how he comes to understanding must be resolved by any rider who would compete successfully. Keith doesn't consciously confront the philosophical issues of animal mind, but his experiences have led him to the understanding that the boundaries of Faktor's mind do not include ethical consideration of Keith Taylor. Lendon has made no systematic study of the horse's perceptual apparatus, but she is keenly aware of what events disturb Last Scene and what she can to do give him pleasure; and while Anne does not trouble herself with the capacity of Eros or Dynamite for reason, she says, "I know how they think."

Trainers accept without question the differences between their minds and the horse's. It is their job to learn their way around another kind of consciousness. They must know how to bring horse thought in line with their own because each of their sports requires the horse to work alone with its rider. The racehorse, the steeplechase horse, and even the polo pony work alongside other horses. But the show jumper, the event horse, and the dressage horse must do what does not come naturally. They must leave the herd and think for themselves.

Horses are intensely social, and the herd, the band of mares with its stallion and outcast adolescent males, is the basis of their society. In the herd, horses stand in close physical proximity. This bunching is their comfort, their protection, and their pleasure. They touch each other constantly, grooming each other with their lips and teeth, brushing away flies with their tails. There are social rankings—the boss mare, the second mare, the young lieutenant stallion—and a horse's status is part of its identity and personality. But a herd is not a pack, and a horse is not so driven by status as a dog, which cannot see another creature without determining through some means which of them has or shall have more status. Friendship is important to a horse, and even though they live in groups, horses fall in love easily. Two horses—even two of the same gender —who happen to be loaded on the same trailer will form a passionate friendship that will reignite any time the two reencounter each other. At any horse show there are always in the background the calls of horse friends bugling urgently to each other.

It is often said that the cruelest thing you can do to a horse is to keep it alone. Even though I believed this, I ignored it once. I had borrowed a friend's horse, a small gray-going-white Arabian, to keep company with my own horse. For years, Wave had earned his living as a companion horse, and like a career diplomat, he had flawless manners and knew how to keep the other party—any party, horse or human—happy. When my mare died, Wave showed no signs of distress, so I decided to keep him until I acquired another horse. My husband was uneasy about the situation and grew more so when I did not bring in another horse immediately and Wave's solitude continued. One summer morning when fog hung so close we couldn't see the barn from the house, there was a plane to catch. I turned Wave hurriedly out to his pasture and left without checking all the gates. As the suitcases were going into the car, I heard hoofbeats on the road. Wave's pale gray rump and white tail vanished into the fog. He was running for the highway, which he believed would lead him to another horse somewhere. He was invisible in the fog. No driver could have spotted him in time to stop. Fortunately, he had experience with cars, and he dodged along the highway for a couple of miles. But he gradually must have realized that a friend would be very far away and become discouraged. He left the highway and circled back through the woods and open farmland to our barn, where he could find something familiar, even if that included solitude.

When Keith rides Faktor down the road, he is taking him away from his own world and asking him to go alone into a new universe, where the only body close to Faktor's is an insubstantial substitute, a human body. When Eros walks into an individual shipping container, which is rolled across the tarmac and elevated to the height of an airliner, he is leaving behind his natural haunts and predilections to fly to Europe, where Anne will show him unfamiliar obstacles and ask him to figure out how to leap over them as fast as he can without touching them. These transactions occur outside nature. They rupture the bonds of the herd and force the horse to exist as an individual. How willing the horse is to enter this new state of being, and how far he is willing to explore it, depend upon the strength of the bonds with which training replaces the natural affinities.

Training is the language of physical response, and this communication is what supports the horse in its activities away from the herd. What allows Last Scene and Eros and Play Me Right to compete at the top of their sports is that they trust, they lean on, their communication with Lendon and Anne and Keith. It has become more important to them than what goes on in the pasture.

Keith was right. Faktor didn't know what he was doing. He also didn't know what he could have been doing, because he wasn't relying enough on Keith. But this takes time. It is unusual for a horse to accept human instruction readily, and the horse that is too docile to exercise its own will probably won't have enough fire or initiative to be good. Although Last Scene's manner has become quite urbane—he never shows the slightest surprise about humans, their machines, or their agenda for him—Lendon described the little gray as feisty.

"He was wild," she said about the Last Scene who arrived at her stable, "really crazy." He did not want to attend to human ideas. Last Scene wanted to run. He was a relative of Seldom Seen, the horse who had done so much to help Lendon make a reputation, and he was Peg Whitehurst's last horse. But he had a very different mentality and temperament than Seldom Seen, and after a month or two of work with Last Scene, Lendon's initial hesitation developed into serious reservation. "I told Peg,

'This just isn't going to work out.' " Last Scene, however, stayed where he was, mostly, I suspect, because Lendon wanted to please Mrs. Whitehurst, whose confidence in Lendon had created so many opportunities.

"What he needed was someone to walk him around and say"—Lendon leaned back slightly to give the body language *halt*—" 'Whoa . . . *good* boy. . . . Whoa . . . *good* boy.' " For two years, Lendon and her working students traded off duty on this simple task, repeating the routine thousands of times before Last Scene accepted its wisdom. "If I'd had Last Scene at the time I started training Seldom Seen," Lendon says with certainty, "I'd have ruined him." Fortunately, Seldom Seen had been "more forgiving, and kind of long-suffering."

If Last Scene's response to a rider's guidance was to run from it, Faktor's was to back off, to avoid moving forward. "He doesn't want to take me by the hand," Keith told me the first time I saw Faktor. He was talking about the way the red horse felt through the reins. It was this reluctance that caused the horse to stand on his hind legs rather than walk past a puddle. At that stage in Faktor's training, it was more comfortable for him to say no than to give credence to Keith. A year and a half later, the massive chestnut horse was moving forward fairly consistently to take Keith's hand.

If it takes Faktor and Last Scene such a long time to learn just the essentials, does that mean that these horses are stupid, or that, as people—even some of my friends who keep horses—say to me, all horses are stupid? That depends on what we mean by stupid. But I suspect that the reason it takes more time to train a horse than it does to train a dog is that the difference between horse consciousness and human consciousness is greater than the difference between dog thought and human thought. For hundreds of generations, dogs have been bred selectively for characteristics that make it relatively easy for them to live in intimate circumstances with humans, and the qualities that make a "good" dog—obeisance and eagerness for human affection—are not valued so much in horses. In fact, even though many horsemen and horsewomen own dogs, they scorn canine characteristics in a horse and apply the term "dog" to a horse pejoratively. When I told Keith that my horses came to my call, he looked a little disgusted. "That is a *dog* trick." I agree with him that most of the quickly learned behavioral routines we ask dogs to perform would belittle a horse's real capacities.

Teaching a horse to jump fences is more like instructing a child to read than like training a dog to stay or to fetch. Certainly much of the first decade of a young human's learning is devoted to following a teacher through the repetitions that will culminate in basic skills. Anne's system for teaching a horse to jump progresses like a workbook that begins with the alphabet, proceeds to words, and then leads to sentence structure—except that the alphabet that leads to the sentence she will eventually diagram begins with a single rail lying on the ground. When the horse can read that pole in relation to his body and motion, she adds a second rail, then she sets the rails on a circle. Eventually, the horse is reading nonobvious distances and relating these to heights and widths in a sequence.

"Highly intelligent." Hoffy said this quite emphatically when we were talking about Eros. Then she said, *"Highly."* But our culture has traditionally assigned humans exclusive footing on the top rung of the ladder of intellect. Descartes's frequently cited views on animals' capacity for reason seemed to have closed the subject to consideration until quite recently. He described reason as a "universal instrument" that allowed human beings to respond to endless changes in circumstance. Language was evidence of the presence of reason in humans. The fact that animals do not have language as we have it "shows not only that animals have less reason than men but that they have none at all."

My eighth-grade science teacher toed this familiar line. He became one of the great disillusionments of my childhood. He was a favorite of mine because he was young, he was a strong batter, and he laughed at the right times. But he lost me when he told the class with great, final authority that animals do not think. He explained that he meant that animals are not capable of reason. I raised my hand to point out that my friend's horse Sandy was a whiz at opening gates.

"He nudges the bar that slides out to fasten the gate to the post? That's because he got lucky once, and the gate opened. If the horse could reason, he would take the lesson of the latch and apply it to some other problem."

I said, "If there's a snap on the gate, he undoes that."

"Luck," said my teacher.

"And if we tie a gate shut, he'll take the knot out and then move the latch."

"He fiddles until he gets lucky. That's all it is," my teacher ruled. "It has to be, because animals can't reason."

The man crumbled before my eyes. It was clear to me that he knew nothing, and it is clear to me now that my own education has been mostly a matter of fiddling until I get lucky. I try things out until they work. But my teacher was just repeating the received wisdom of the time. This held that only the most highly evolved animals, meaning humans and possibly some primates, could "think." Beyond that *he* did not want to think.

Descartes was actually somewhat more open-minded on the subject of animal thought than my teacher. He construed the intellectual activity of animals, taken as a group, to be a mildly haywire variety of human consciousness: "Animals do not see as we do when we are aware that we see, but only as we see when our mind is elsewhere." They are like us, only not quite as sharp. This notion seems to go hand in hand with those that equate the minds of children with the minds of animals. But I think there has always been a popular corollary suspicion that animals have greater intellectual powers than we can divine. Descartes himself, in personal correspondence, allows for this undetectable form of mind: "Though I regard it as established that we cannot demonstrate that there is any thought in animals, I do not hereby think it is demonstrated that there is not, since the human mind does not reach into their hearts."

Early in this century, our suspicions about the cognitive powers of animals led to great public interest in the case of a German horse named Clever Hans. Nearly every day, Hans and his owner, Herr Von Osten, would appear in a courtyard near Berlin to demonstrate to the public, free of charge, Hans's ability to answer questions about mathematics, harmony, and history and to indicate the location of objects. Hans responded to questions by tapping his right hoof and by pointing his muzzle to indicate symbols on a "writing board." The horse's accuracy was phenomenal, and scientists, as well as the general public, hurried to investigate. Von Osten's lack of interest in commercial exploitation and his willingness to cooperate with even the most formal investigation of Hans's abilities by an academic delegation called the September Commission seemed to indicate the sincerity of his efforts with Hans. At first, the scientists could find nothing but a horse who was very good at arithmetic. They were looking at Hans. A later investigator shifted the focus to Von Osten and found that, apparently unaware of it himself, Von Osten was giving off minute physical cues to guide Hans to the correct answers. The horse had simply learned to read his handler with great accuracy. Hans died in ignominy, the inspiration for the term "Clever Hans fallacy," which has been used to deride a number of behavioral experiments that attempted to show that animals comprehend language as opposed to physical signals. Although cleared of fraud, Herr Von Osten was greatly disappointed in his horse and retired Hans from public view.

My own reaction to the accounts of Hans's exploits is that Von Osten had one exceptional horse. Hans was not only very clever at divining what was on a human's mind, he had the desire to respond to Von Osten. To Hans, the operations of mathematics were probably irrelevant, while the engagement of Von Osten was most important. Enlisting Hans to tend to chores that arise solely from the human intellect seems a novel but marginal test of horse intelligence. What would the hyperresponsive Hans have been able to achieve with a rider like Lendon Gray, who seeks *horse* potential so far as the limitations of the individual horse will allow?

The arithmetic that belongs properly to the horse is the numbering of strides, their speed and rates of compression and expansion. Clever Eros. He can fit in a few more or throw one away. Anne Kursinski may have had the approximate answer before she posed the problem of the big fences at Luxembourg, but Eros knew how to balance the equation. He estimated the distance and the height, divided by the length of his body, and factored in Kursinski's weight. This is a little too cute and much too slow for what went on in the arena at Luxembourg, but the test of height and width and the optical illusions of the fence arrangements and paint jobs are a more valid way to test the intelligence of a horse than Clever Hans's automatic tapping on cobblestones to respond to questions that test human problem solving.

Even though it seems obvious to Anne and to Hoffy that when Eros trots into a grand prix arena, he sees the pattern of bright uprights and rails as a set of questions, a riddle to be solved by a series of efforts he will make, it is not so readily clear to biologists and ethologists and psychologists. They have yet to resolve to their satisfaction the central question about animal minds: are animals conscious? Many scientists still refuse the possibility that animals experience some form of consciousness, while scientists arguing for the existence of consciousness in species other than the human find evidence of it even in species as low on the evolutionary totem pole as the honeybee. I suspect that Anne and Keith and Lendon would find this debate the conversation of educated fools. Whether or not planning the approach to the next fence or making the effort to travel bent in an arc constitutes consciousness, every time they are saddled, Eros and Faktor and Last Scene must begin to consider and account for objects and events.

Nothing is closer to Eros or to any animal than his own body. Whatever degree of consciousness he possesses, it is more directly and far more intensely linked to his body than our consciousness is to our bodies. He must protect this body, and fear is the unconscious subcurrent that can always erupt to cause the horse to do something unthinking, to bolt or kick or rear. Horses are frightened by many things, some of them the same things that frighten us—loud noises, pain, entrapment, falling. But because they are still in touch with their experiences as quarry, they are deeply suspicious about many more subtle events—a new raincoat on a familiar person, a slight change in the flavor of their drinking water, a faraway sound. Certain places, a clump of reeds, the corner of a paddock, can become the objects of great superstition, to be avoided at all costs. Even wind can distract a horse by masking the other sounds the horse has come to rely on. Will James describes a horse's spontaneous reaction to this kind of fear in his account of Smoky's introduction to the saddle: "That hunk of leather was drawing all his interest and ears pointed straight at it, eyes a-shining, he snorted his suspicions and dislike for the looks of the contraption laying there waiting it seemed like to jump up at him and eat him alive."

One of the horses I have most respected was a blue roan mare who stayed with us for a number of years. At the time, I was in awe of her. Magic was aloof and slightly arrogant, and she had an uncanny ability to get things right. She was a good jumper who knew her job the way Eros and Play Me know theirs. She jumped any obstacle put in front of her so long as the rider was not trying to hold her to some human scheme for how to get to the fence and take off. She could not tolerate having the movement of her head restricted. It was claustrophobia, and this was the reason she was terrified of being tied or even cross-tied with lines snapped to the sides of her halter. It brought on panic, rearing, scrambling, thrashing, broken halters and leads, and a lot of dust. But Magic would consent to the ground-tie. She would respect a loose rope hanging from her halter to the ground as if it were knotted around a telephone pole. It took only a few days for us to recognize and adapt to her fears, and shortly thereafter we took Magic to horse trials where we encountered the person who owned the stable she had come from. Our friend Frank Page was cavalry. He had ridden under the renowned Harry Chamberlain at Fort Riley, Kansas, and he had endured the ill-fated World War II mission to deliver horses and mules over the Burma Road to Chiang Kai-shek. He had coached generations of Cornell men in polo, and the bumping in the game had left him with one stiff knee. But at nearly eighty, Frank was still remarkably strong, and the hitch in his gait was the only compromise in his fitness. If you kept a horse in Frank's stable, you followed his way, the cavalry way, or you took the horse down the road. Frank saw Magic standing ground-tied as we were saddling her. He limped over to me and put his face close to mine, because this was the way he always spoke to anyone. Sometimes he remembered to remove his cigar.

"What the hell is *this?*" he said about the rope dangling to the ground.

"She doesn't tie," I explained. I was certain he must know it.

"The hell she doesn't! I never had a hoss in my barn wouldn't tie!"

"Really, Frank. She doesn't tie."

"You bring the hoss to my place," he instructed. He wasn't one to be crossed, and he had helped us so often that I wouldn't risk offending him. But I unloaded Magic at his place with dread. She wasn't my horse. She wasn't his horse, and her fear was so great that when she found her head caught, she would fight until she hurt herself. Frank took her from me. "See ya later."

Several days later I returned to find Magic calm and happy in the stall where she had always lived. On each side of her halter was a loop of baling twine.

"Frank, what are the strings for?"

His face pushed up close and belligerent, and the cigar shifted roguishly. "What strings?"

Accommodation had been reached. The string loops were panic protection. They would snap as soon as the mare began to put up a serious fight. Magic could remain a horse who only ground-tied, and Frank could remain a horseman whose horses always tied.

Often a horse can learn to think through things that frighten him, and this is one of the tasks every rider faces—how to give the horse the confidence to respond to environmental challenges without panic. Some horses, like Eros and Play Me, are better at this than others, braver and bolder and able to make more independent judgments. But for all horses, confidence is the essential lesson.

Even when it is not threatened, Eros's body is, in a sense, always on his mind. It gives him place and status. In his stall between rounds at Lake Placid, Eros stood with his small refined head above the half-gate that closed him in. The slight dish between his large eyes made them appear to pop a little. He was waiting for action. His lips fluttered slowly at the edge of the half-gate, his teeth occasionally taking hold of the metal, just venting a little energy. His engine was idling. There wasn't much on his mind until the corgi that travels with the stable, a low-slung little dog with herding instincts, sneaked under the gate and began stalking Eros to force him away from the stall door. Anne was seated just around the corner. Horses at pasture will often chase and attack dogs that trespass on their grazing, and Eros had ample opportunity to kill this little dog with a blow to the head. But instead the horse went through channels. He immediately kicked out behind—not at the little dog—at the wall. He achieved the desired result. Anne stopped talking and called out, "Does he have hay? What is his problem?" Eros let fly another hind foot, struck the wall, and his problem was solved. Hoffy removed the dog from his presence, and Eros took his place back at the gate.

Clever Eros. He uses his body to sort through experience and to create strategy. If he tips a rail out of its brackets with a hind hoof, he understands this as an error the way we understand stumbling as a blunder. If he hits the rail with his leg, it smarts. It's the same kind of wake-up call you get when you stub your toe. Usually he's a quick study, and even as he moves ahead at top speed, he can calculate how to correct the problem a couple of seconds later at the next fence.

If Eros were to actually fall, the experience would distress him. Horses are at ease when they are poised on all fours and in a position to take flight. A horse that goes down is subject to predators and is powerless. Falling is frightening and humiliating. It is deathy. Old-time horse trainers used to exploit the effect falling had on a horse with a device called a throwing harness. Various straps and buckles and rings that linked the horse's head to its legs enabled the horse's driver to bring the animal suddenly to its knees and hold it there. The idea was to tempt the horse to misbehave, then force it to fall. This was the theory applied to the rearer I had as a teenager, The Candy Kid. The old man who stabled my horse at his farm began to hear reports from the other kids that the chocolate-colored gelding was rearing and throwing himself over backward, and he noticed I had stopped riding the horse in favor of his own horse, Sandy. He had farmed with horses, delivered the mail with horses, traded horses with the Amish, and to him the horse's true work was to draw a vehicle. He had a throwing harness and a friend a little younger who had assisted before. On a day when school was let out much earlier than the two of them expected, I came down the lane to find the men driving The Candy Kid to a two-wheeled training cart. There were several ropes attached to various bits of harness. They walked along with the horse, making him move forward with a long carriage whip. Every time he showed any inclination to hop or lift up, his nose would suddenly meet the grass. He would be held there kneeling for several minutes, then asked to get up and move ahead. It was surprisingly quiet, with no apparent struggle, and when the horse pulled the cart back into the yard, it was difficult for me at the age of thirteen to perceive any difference in him. We would have to see, the men said. They were planning another disciplinary session, but they were never able to put the throwing harness back on The Candy Kid.

A couple of weeks later, the horse went down the way horses fear they will. Poisoned by something he grazed on in pasture, he died a violent, barn-rattling death. He was in great pain that kept him down in his stall, rolling up against the walls, intermittently struggling quickly to his feet as if he were alarmed to find himself on the ground. I think of him when I see a horse take a fall at a fence and instantly scramble to get up. He fought to remain standing, but the intervals for which he could endure this became briefer. His death throes went on too long, I realize now. The veterinarian who was standing by should have removed him from the agony, but finally, hours after his ordeal had begun, the horse fell down and lay still.

A horse's death is momentous and terrible. The animal

is so large that there is no way to fall except hard, and once the horse goes down, he is powerless to express anything. I think that is why falling is the most cataclysmic in a range of failures that upset a horse. Going down can cause it to regress. "Don't teach the horse what he *can't* do," Anne says about riding at big fences too fast and too soon. The trick is to increase the height and speed and complexity without fracturing the horse's confidence.

You might watch Eros jump and say rightfully, "Clever Anne." But in fact even the best rider has only limited means to make what the horse learns about performance important to him. Jimmy Williams's background in stunt work had made him open to whatever solutions were productive. "We tried everything"—Anne smiles ruefully—"electricity—everything," and there are still pieces of horse hardware named after Williams. But watching his experimentation caused her eventually to reject most of the innovations and return to the means of classical horsemanship. There are only a few tools in this kit: a bit or other device to control the horse's head through the reins, the saddle, and the rider's body. Anne can use her legs, her back, her weight, and her voice. After that, she has the so-called artificial aids, the whip and the spur. Centuries of use in metaphors of cruelty and exploitation have caused wide misunderstanding of the whip and the spur. The horse is an extremely large animal and heavily muscled. The whip and the spur are tools to extend the emphasis of the rider's leg against the body. The use of either involves technique, and while a whip or a pair of spurs embodies numerous possibilities for abuse, the knowledgeable rider recognizes that there are many situations in which a horse doesn't have to be dead to make a beating futile. The trick of the whip is to learn how to touch each particular horse to get the right response. Repetition of this response is the way the horse comes to understand the rider—and vice versa.

Keith's style with a whip is light and quick. With Faktor, the whip is often active, steadily flicking the heavy red sides behind Keith's leg.

"I think he gets bored," Keith told me. "I can feel him kind of blink off."

The whip is not painful to Faktor, just irritating. When his motion is the one Keith wants, the whip comes off and the irritation is gone. When Faktor's mind begins to wander . . . tap, tap . . . a reminder that he still has work to do.

As I watched each of the riders at work, it became apparent that there is always a tension between boredom and routine. Boredom can dull a horse, routine can soothe him. In natural settings where horses graze together at will, they make paths, narrow wiggly little paths like the ones cows make. These are safe paths the horses can take in a head-to-tail torpor in their daily wanderings, which are timed by patterns of feeding and drinking. The paths and the patterns comfort horses. They reinforce a horse's sense of well-being. The same is true of patterns in handling and movements in training. Many horses seem to lean on routine. Anne told me that both Eros and Ferina appeared to take great support from what she calls her program. But she also tells students, "Do something *different*. Don't keep doing the same thing," and I hear exactly the same warning from Lendon. Repetition is soporific. Change in directions, degree, pressure, speed keep Faktor's attention without disturbing him.

Sometimes even a sensitive rider can fail to keep the work engaging. It becomes tedium, and it is deadening. The novelist Tom McGuane, who spent a good deal of time around horses trained to cut cattle, tells of a cutting horse stallion who abruptly lost not only his edge but his willingness to even be led by hand into an arena where cattle waited to be separated. "Sumbuck's got so much cow," the trainer told McGuane, "he can't go in there and face it."

But to Faktor, the tasks themselves are still interesting. Two years after I had first seen him, the big red horse had come a long way. He was quite lean and fit, and his work in dressage had progressed to lateral movements. Keith said, "Watch how clever he is at this," and the horse moved ahead passing sideways at the same time. The tricky cross-steps looked easy. Clever Faktor. He could make the moves now. He was galloping faster, jumping bigger fences. Since Keith's injury with him in the spring, Faktor had moved up through the lower divisions, winning fairly consistently at Training level, and Keith was readying him for

Preliminary-level horse trials in Virginia. While Preliminary may seem to connote a beginners' competition, in combined training it means the first level at which the horse begins to earn points. It is Preliminary to the Advanced level, where the horses must be exceptional athletes. Preliminary is the level at which rider decisions and rider mistakes become important and where many amateur riders stop, either because of their own limitations or because their horse cannot get around the jump courses. Even at Preliminary, the horse must be a very good one.

Faktor went to Morven Park, where the horse trials were actually offering a purse—"real money," Keith joked, but he wanted it. He and Faktor earned the best dressage score. They were still winning after running the cross-country course. Then?

"Three rails," Keith summarized their performance in the stadium phase.

"*Three?* What happened?"

"Three."

Second place, no prize money.

Faktor had figured out that the decorative show jumping fences were somehow different from the natural obstacles on the cross-country course. They would topple, and bumping the rails out of their cups couldn't hurt that much. What did it matter to Faktor if he left a few colored poles on the ground? Combined training is an unforgiving sport. To win, a horse must excel at all three phases. Faktor wanted to coast through the third one, and it's certainly not unusual to see placings come crashing down with the rails of the stadium fences. From the horse's standpoint, lapsing in the final phase is partly a matter of just not having enough stamina left: *I controlled myself in the dressage and gave you a good test. I attacked your cross-country fences. I ran. I jumped. I came out clean. And so? I came out tired. Now you want me to make the turns, make the heights, make the time, and be* careful?

"We're going to be doing a lot of show jumping this winter." Keith's plan to sharpen Faktor's jumping style was also aimed at his own skills. He suspected part of Faktor's carelessness originated with him. Show jumping calls for intense focus from both the horse and the rider, and photos of Keith that appeared in one of the horse magazines illustrate this focus wonderfully. But remembering Keith's remark about how Faktor became bored and blinked off in training, I wondered if Keith reinforced Faktor's boredom because their work routine at that moment bored him too. I wondered if Keith was having last-phase lapse himself. He's an able and tactful dressage rider, and he understands the strategic importance of the phase. His form over fences is classic. He's smart about lines and turns, and in his stadium rounds he tries for rhythm and smoothness. But after the speed and danger of the cross-country course, maybe even competitive urges at the prospect of actually winning weren't stimulus enough to protect Keith himself from being touched by boredom during the comparatively civilized progress of a show jumping round. Maybe Keith and his horse blinked off at the same time.

Because the horse moves in such physical intimacy with the rider, the human's attitudes and sensitivities are transmitted readily to the animal—and vice versa. I was amused to see that Lendon's approach to learning to ride was mirrored by Last Scene's learning of the grand prix movements. When Lendon was studying with Michael Poulin, she spent hours wrapped in horse blankets, sitting in the frigid indoor school to observe Poulin's other students so that when it was her turn in the saddle, she could get it right, exactly right.

"If Michael asked me for more bend in the horse, I could deal with that. But if he had to ask me a second time, I'd be furious—not with him, with *myself*."

This mix of fury and generosity is evident in the way Last Scene absorbs teaching. In the early spring of 1994, Lendon took the horse to a qualifying competition for the World Equestrian Games that would take place that summer in The Hague. "He's small, but . . . what the hell?" The horses performed individually in a vast new arena in Lexington, Virginia, where every detail, including sound, had been thought through by engineers. The place was light, very clean, and in spite of a distant backdrop of relentless New Age music, incredibly quiet. The arena was large enough to accommodate two dressage rings, but we could still hear the horses breathing as they carried out the movements of their tests. Set in such enormous space, even the largest horses looked small, and paradoxically, Last Scene's size disadvantage was hardly noticeable. His surroundings leveled the playing field somewhat, and the horse came into his test with the expression that is typical of his concentration. His ears were swiveled back. He was trying. Each of the movements was done, and although the score Lendon received qualified the pair for the final selection trials in June, the horse accomplished the movements and nothing more. Last Scene knew how to skip along, changing his leading legs at each stride of the canter. He knew how to do pirouettes and *passage* and *piaffe*. But he did not yet know how to make these moves his own moves.

The movements required for a grand prix dressage test and the transitions between them are so difficult that simply enacting them proves the horse's extreme effort and cooperation. The *piaffe* is often a case in point. It is always hoped for but never expected as a matter of course. The first one comes after several long, intricately divided minutes of other movements to begin the climax of the grand prix test. If the horse cannot lift each leg to enact the trot sequence at the standstill, this is a big letdown, and when there is more shuffling or sometimes even a complete holdup at the next *piaffe*, there is more disappointment.

A year or so after Last Scene's qualifying test in Virginia, at the U.S. Equestrian Team Festival of Champions,

Carol Lavell rode Gifted, the horse that took her to a bronze medal in the Barcelona Games, in a grand prix competition that would determine a national championship. The festival is held at "the Team," the USET headquarters in Gladstone, New Jersey. The Team is an expansive estate originally built by James Cox Brady. There is enough land to house cross-country courses and the marathon phase for carriage competitions. The main stable is a huge brick building from the nineteenth century, more substantial than most homes from the same era. The wings where the horses are kept meet in a round, two-story atrium under a cupola. The stable floors are brick, and the bars separating the stalls and the finials on the corner posts are brass. Behind the stable, there is a large sand arena enclosed by stone walls. The judges' booths have Plexiglas walls so as to obstruct no one's view of the horse at work in the dressage ring, and there is a large digital scoreboard to provide nearly immediate readouts of the judges' assessments. The stands beside the arena were filled beyond capacity. The man next to me apologized for squeezing in to occupy part of my lap, but he said he had seen Gifted once, and he had to see the horse again.

Gifted is an eyeful. He is immense, a brightly marked dark bay whose every move is big. His stride has amplitude and freedom and precision. His trot shakes the ground and he goes forward with an expression of great integrity.

Dressage spectators are the most decorous of the horse sport fans. They respond politely to all rides. They appreciate the effort required of the horse, and, respectful of this, they maintain silence. But people love Gifted. Each horse had to pass close to the stands on his way to and from his test, and as Gifted thundered toward the spectators, there was spontaneous applause. After entering the ring, the long sequence of his movements went well, impressively, in fact, until the first *piaffe*. Gifted did not begin the movement, and he wore an absentminded look, as if *piaffe* had just slipped his mind. Because the competition was a USET championship, the riders were not permitted to carry whips, which are often seen in dressage shows. During a test, riders cannot apply the whip, but its presence reinforces their requests to the horse. Without a whip, the only means Lavell had to bring Gifted to *piaffe* were her seat and legs and spurs. The second occasion for *piaffe* passed with an unenthusiastic lifting of a front knee, and at the last, climactic *piaffe*, the horse appeared to draw himself together for the effort and then let down. Lavell rode out of the arena toward the spectators, who, in spite of the absence of *piaffe*, stood and applauded and called out. Lavell has spent years on this horse's *piaffe*, and she rode past us facing straight ahead, her face composed with a smile. She looked ready to weep.

The next day, when the pair were to compete in the musical freestyle, I returned with a friend who frequently judges lower-level dressage. There was no championship at stake, and Lavell carried the slim black dressage whip. The spectators were quieter, waiting to see what Gifted might show them. The music Lavell had chosen was as bold and broad as the horse. I wondered if she was dreading the *piaffe*—nothing like a letdown accompanied by momentous music. As the horse strode toward the arena and the music began, Lavell flicked the whip lightly against the horse, and once in the ring, Gifted put his mind to business. When he made the first *piaffe*, there was whispering all around us, relief among the fans, and when he moved off from the second *piaffe*, I strayed from protocol to whisper to my friend, "He's dancing today, isn't he?" She nodded at Lavell's whip. "Hummin' too."

Gifted raised his legs in order and sustained the cadence. This is all that is necessary to satisfy the extreme demands of the test. The horse that adds any element of verve to this motion is offering a gift.

Acquiring élan takes time and the gradual accrual of physical strength. In March, Last Scene's score qualified him to compete in the trials to select the U.S. team for the World Equestrian Games, and when June came around, I was not surprised to see Lendon's hardworking little horse put in much the same kind of test. He had not had enough time to learn how to put more into moving or to develop the strength to do that. This began to change later in the season.

Early one autumn morning I stepped into the gloom of the indoor school in Bedford. Only one rider worked there. It was Lendon on Last Scene. The movement was *piaffe*. She didn't say hi. She didn't look up. She said, "Please keep moving. You're blocking my view."

I was standing in front of a tall mirror that allowed Lendon to see Last Scene's movement without changing the position of her own body. She was clucking, using her legs, reinforcing their pressure with taps of the whip, and watching. They moved in and out of *piaffe* several times. Last Scene hesitated somewhat. Then he started the movement snappily but bobbled the rhythm. They moved closer to the mirror, Last Scene approaching his dancing double, and the movement came together. The cadence was correct. There was elevation, nearly equal elevation of all four legs, and there was something else too, something like exuberance.

The next time I saw Last Scene he was dancing for a crowd. Music filled the big arena at Port Jervis, New York, and there were two horses dancing a *pas de deux*. This is something often performed for dressage audiences, but today's was a highly unconventional one. Last Scene, mane done up against his neck in formal braids, carried Lendon in her top hat and tails. The other horse was a red chestnut stallion about the same stature as Last Scene. The chestnut's mane was free. He wore a Western saddle and carried a cowboy in an outsize Western hat. Sonny's Cash Lander was a national champion reining horse, and his moves were movements from the Western tradition of horsemanship that were roughly equivalent to the movements from the

European tradition. While the little stallion spun on his hindquarters, Last Scene marched up to him and made *piaffe*, and more *piaffe*. It looked effortless. It went on and on. The dance climaxed in a long tango chase in which Sonny's Cash Lender backed straight across the considerable length of the arena, pursued by Last Scene in a hundred bold, vehement strides of *passage*. It was the longest interval of that movement any of us in the audience would probably ever see. The spectators hollered and stamped and whistled. "This is a very generous horse," Lendon told the crowd, and the little gray horse struck out across the arena, extending his trot, firing his legs straight out from his shoulders and hindquarters. Last Scene had the moves now. He used to know them. Now he owned them.

One of the arguments that are sometimes used to deny the possibility that animals other than humans have consciousness is the speculation that animals lack self-consciousness—that is, they cannot perceive their own activities objectively and they lack awareness that they can be perceived. But Last Scene could not have performed his dance at Port Jervis without monitoring his movement or apprehending in some way that he was being watched.

Lendon shared with me an experience that made it clear that horses make use of self-consciousness to learn. Quite early in her career, she worked with an Appaloosa horse that she would eventually take to the coaching sessions for the USET combined training team. Pride could jump and gallop, but he had difficulty in the dressage phase because he would not or could not extend his trot as Last Scene did so brilliantly. In spite of Lendon's persistent work, Pride could not be made to move differently, and the problem was not resolved until the horse resolved it himself. Lendon and a friend were standing outside a paddock where Pride was turned out with other horses. There were horse games going on, and the girls watched Pride trotting about with his friends. Then, in a moment of horse play, Pride thrust his front leg forward. "A light came on," Lendon says, "and what was remarkable is that I could see it come on." Pride recognized his random movement as something Lendon had been looking for, and as soon as he did, he was able to use this new knowledge when he was carrying a rider.

Last Scene, living up to his theatrical name, now recognizes an audience as an occasion, a call to give more to movement. Undoubtedly, some of this anticipation is transmitted by Lendon, who relishes performance herself, but this is only a partial explanation. At a small show in the Hudson Valley, the spectators were sparse. After riding a number of horses, Lendon stepped up into Last Scene's saddle and looked around. "Where's my screaming crowd?" She laughed. From the warm-up ring on top of the hill, you could look down on the grass dressage arena. In a few minutes, the competitors riding to music would begin. During the warm-up, the little gray worked manfully up through the movements of increasing difficulty, ears swiveled back in concentration. He made *passage*. He made *piaffe*. Then the music for the first freestyle test came over the public-address system. Sounds of the orchestra seemed to lift Last Scene. The *passage* became more elevated, the *piaffe* crisper. When Lendon let the horse pause to catch his breath, Last Scene looked around for the source of the music. His large dark eyes settled on the competition arena with its abbreviated grandstand and little clutch of spectators. He raised his head, and his ears came sharply forward. He moved off again with Lendon in a good-natured little fury. *You want* piaffe? *I'll show you* piaffe!

He had not yet entered the ring where he would perform. He had not yet even passed close to his spectators. But—conscious or not, self-conscious or not—Last Scene knew he was there to make his moves, and he knew the rest of us were there to watch him.

Fear and Falling

From the side of a big hill overlooking a large artificial pond, we can see Play Me Right galloping toward the water. Keith is straightening, sitting back a little to gather the horse for the jump. Between the horse and the water is a short, sharply upward run to a huge log poised at the edge of the water. This is the Fair Hill International one year after Play Me Right had been pulled out of the competition before the start of the endurance phase. He is back and he is running, into the shadow of the bank. As he approaches the big log, all we can see of Keith and the horse are the tips of the horse's dark ears. By the time the bay head appears, it is clear that Play Me is "looking at" the fence. Something about it, the huge log after an uphill run or maybe the horse's first glimpse of the water on the other side of the log, is worrying him. For a split second, the old horse loses his gameness. He seems to shrink back into himself and hesitate. Keith uses the whip, and Play Me goes for it. He leaps up over the log. As he passes over it, a front leg catches and brings him down on the log. The horse is halfway through his dive into the water, and he is stuck there.

Six weeks earlier on the course at Morven Park in Virginia, Play Me Right came over a similar log and dropped into the water, where he landed wrong and fell. The fall hurt. He was laid up with an injured foreleg. This injury had itself been foreshadowed in August by a misstep on landing in the water at the horse trials in Millbrook, New York, in which Play Me stumbled violently enough to throw Keith off. Perhaps the accumulation of difficulties he had experienced jumping down into water was enough to make Play Me hesitate.

But these mishaps over the summer were just that and not enough to mar an otherwise very successful season for Play Me Right. As a three-star international three-day event, Fair Hill is as close to Olympic-level competition as the Americans get. The British, who, along with the Australians and New Zealanders, have dominated this sport, host two four-star events, and there is a good deal of talk in the combined training community here about how difficult it is for American riders to gain experience at this level and how little it would take to build the Fair Hill course up to the four-star level and provide home-turf training for the Americans. In the meantime, the two home courses to tackle are still the Rolex in Lexington, Kentucky, which is run in the spring, and Fair Hill, which climaxes the combined training season in this country. With Faktor becoming competitive in the lower divisions of horse trials along the East Coast, Play Me ran well enough in the advanced divisions to qualify for the test at Fair Hill. Fitting him for the course was part of the test.

Because of the foreleg Play Me injured at Morven Park, Keith stayed off the horse's back until two weeks before Fair Hill. It was back to the swimming pool to keep up Play Me's cardiovascular fitness, and this kept Keith busy. It was a half day's work to wrap the horse's legs, drive down through West Chester, unload, unwrap, and then work back through the process in reverse after the horse had swum for a half hour. The pool was indoors, a moat surrounding a concrete island populated with tropical plants and guarded by statuary greyhounds. Play Me's swimming was supervised by the pool owner, who was very proud of the pool, how clean it was, and the size of the pumps and filters. He fitted a pole to Play Me's halter, led the horse down the ramped floor into the water, stepped across a plank bridge, and began to walk around the plants while Play Me swam around the concrete island on the pole. The pool owner told me it was usually just this simple. He would lead the horse down the ramp. The horse would walk slowly until the water was too deep to move this way efficiently and then would begin to swim. He said the horses never fought it, and the only horse who hadn't taken to it right away didn't seem to recognize that he could swim. "You know, he just kept walking and pretty soon he was blowing bubbles like some kind of hippopotamus!" As a beginner, the sinking horse wore a safety line on his tail. Tugging on this line elevated his hindquarters, and he was swimming in spite of himself.

Play Me Right didn't need assistance. He was familiar with the exercise routine, and after each set of laps, he walked up the ramp leading from the pool with his sides heaving. Keith would confer with the man with the pole. "You think he's had enough?" and they would transpose the swimming yardage into galloping miles. Then maybe Play Me would be led back down into the pool for a few more laps.

"What's his distance going to be down there?"

"About ten miles, before he gets to jump—"

The pool owner turned to where Play Me stood with little streams of water running off him and shook his head.

"That's why I like the races, Keith, the flats. Like I told you, it's simple. None of the jumping business. You take the horse, run him, and if he's fast, you get some money. Now here you're going through all this, then you get down there, jump around all those fences, and even if you win, what are you going to get?"

After all the swimming and vet calls, chiropractics, and farrier visits, Play Me Right had passed the veterinarians' scrutiny at Fair Hill. The routines of the horse's physical conditioning allow a combined training rider to make parallel mental preparations—the more you know about your horse's strength and stamina, the better you know the capacities you can draw on. Repeating an exercise sequence for the horse is a way of repeating to yourself the tests that are ahead. Keith had not shown any evidence of nervousness or even anticipation at other competitions, but, perhaps because he had gone through so much to get Play Me to Fair Hill, he was visibly tense riding into the dressage ring. Play Me never made good scores in this phase, and sometimes his penalties were among the highest. All Keith wanted were decent scores, to put him in the top part of the field, and once he was there, the other Play Me Right, the cross-country machine, would take over.

In top hat and tails, Keith looked smaller and somehow younger than he did in his polo shirt and helmet. Once the salute was over and he was actually riding, Keith loosened up. Play Me put in the best dressage test of his career, which left him in the middle of the field. He survived the second vet check and was allowed to start the endurance phase the following day. The horses steamed into the vet box after eight miles of trotting and two miles of steeplechase. It is not a box, just an area cordoned off where the horses are held for ten minutes to be cooled down with water and ice and checked out by the veterinarians before they start their cross-country runs. It is always crowded and busy. Play Me Right came in trotting and blowing. Keith's girlfriend, Sarah Mallas, was waiting for the horse there. She is tiny with long dark hair and strong opinions, and she knows about horses. She knows how to braid manes and how to turn out horses for formal presentation. She knows what needs to be done at the vet check. Almost before Keith could hop off, Sarah and another friend began throwing water over the horse. The hold seemed brief and the grooms harried, but Keith told me this break in the action is always too long for him. "I just want to get started."

The course is a loop about three and a half miles long, punctuated by hills, woods, gullies, and water—and by twenty-three very large fences. There are usually two or three horses running on course at any given time, and spectators scatter along the route. You can walk along with the horse traffic, making your way from one fence to the next or from one prime vantage to another, so long as you respond to the whistles that sound to clear spectators away from a fence as a horse approaches and, out in the galloping stretches of the course, to the calls of "Heads up!" that alert you to step back from the horse's path. Because there is a public-address system and radio operators stationed around the course to inform the announcer, you can stand anywhere on the course and know what's going on with each horse. We chose the hill over the pond and dike called the Chesapeake Crossing, the second water obstacle on the course, because it afforded a view of several fences at once. In fact, it was the point on the course where the horses passed closest to each other. Traffic on the cross-country course is strictly one-way. The horses run alone, and they would be alarmed by the sight of another horse galloping toward them. From the slope, we could see the big log on the pond dike, and by taking a dash over the crest of the hill, we could also watch the horses at one of the last obstacles on the course, a sunken road, where the horses jumped first up to a knoll cut through by a dirt road about four feet below, then down into the road, and finally they sprang up the opposite wall of the road and out to a long stretch of galloping. To our right in a stand of old pines was the Waterloo Rails, a log obstacle in the shape of a narrow corral. This pen came up early in the course, and it was built on a steep grade. The distance between the tall log walls was so short that the horse running down-hill and clearing the first wall of the pen would need to "bounce" the second—that is, land and take off again in the sequence of a single stride. The weather was bright, the footing was good, and Keith and Play Me Right hopped in and out of the pen without taking much notice of the big log walls and kept galloping. At the same obstacle the year before, extreme weather had created the setting for a different kind of challenge and a remarkable occurrence.

Heavy rains had been falling since before dawn. It was late in the day, and any security in the footing had washed away. Phyllis Dawson, a member of the Olympic squad for the Seoul games, galloped over the top of a hill toward the big log pen. Her horse, prophetically named Half Magic, jumped down into the pen, landed in the slime there and skidded, slamming into the second log wall with its chest.

Dawson was shaken loose and lost her reins but was not thrown, and apparently she was resigned to having to circle back and retake the entire obstacle. In the moment she tried to collect herself, Half Magic drew back sharply on his haunches and rocketed up over the logs. He landed running. Somehow Dawson stayed with him, but she had no reins. The horse was galloping, heading toward some undesignated next fence, and he slowed up only when his way was blocked by a line of spectators. Dawson gathered her reins, laughing and shaking her head, and put Half Magic back on course.

"Keep riding as long as you have a horse to ride," is the directive of another Olympic veteran, Mike Plumb. This is the equivalent of Satchel Paige's "Don't look back, something might be gaining on you." The idea is to ride ahead of bad luck and fear. It's a good idea and it works until whatever is gaining on you draws up even and you have to look it in the eye.

We don't talk much anymore about courage, at least in public, and "brave" is a word that comes up most often in stories for children. The reason that most of us don't speak of "courage" and "bravery" is that we rarely call on these qualities and we're afraid of sounding corny. In an article on schooling for jumping, Bruce Davidson, currently the most successful event rider in this country, lightly admonishes neophytes to "be brave," but that is the only mention of this sort I've seen or heard from event riders. Even though these riders do not talk any more about courage than we do, show jump course designer Steve Stephens says, "Now, your three-day riders—now *those* guys —it's hip, hop, and then they jump off a goddamn cliff!" Humans and horses have figured out how to perform some remarkable feats together, and at Fair Hill I realized that the deeds of real-world riders and horses in combined training and show jumping are too phenomenal to make them believable in fiction, especially in the horse stories I was so obsessed with as a child. What happens at big fences in tight places is dangerous and often heroic, like what happens in *Thunderhead* and *The Island Stallion Races*. But it is real, like war, and the accounts you hear about real horses running real courses have the ring of war stories.

Combined training and show jumping require bravery because they are essentially war games, the legacies of a long military tradition in which the cavalry schools of France, Russia, and England provided systematic education of horses and soldiers and lively theoretical debate on horse mastery. In the late eighteenth and early nineteenth centuries military riding masters like the Earl of Pembroke took notice of the success of the Hussars and Magyars, who rode "short" and poised forward, and they began to advocate this for all military men. These methods of managing war mounts across country and over fences remained somewhat separate from the dressage methods taught as manège until the late nineteenth century, when James Fillis, an Englishman who was chief instructor at the Russian cavalry school, began to adapt dressage technique and theory to cross-country riding. Then in the early twentieth century the Italian Federico Caprilli, arguing for the "forward impulse," observed that when the horse jumps and when he gallops, his center of balance is constantly shifting. From this Caprilli deduced that the most effective rider is the one who interferes with the horse's balancing and rebalancing as little as possible: "Tell the horse what to do, then conform to the horse's way of doing it." His "Italian style" still exerts the strongest influence in current ideas about riding over fences.

While the masters were theorizing and publishing, they were also producing riders who in practical effect were professionals. During lulls in military action they honed their skills in competition. The three phases of combined training summarized their education on all fronts, and show jumping, which was easier to stage on an impromptu basis, brought cross-country riding into a game format that could be viewed in the riding school. Military control of these sports, and of dressage as well, continued through the first half of the twentieth century. The teams that competed internationally were made up of officers, and although the American officers made a good showing at the 1932 Olympics with a team that won the gold medal in combined training and Hiram Tuttle's bronze in dressage, it was not until well after World War II that the Americans began to pose any threat to the Germans, Dutch, Scandinavians, or British who had mounted the winning Olympic teams in all three sports.

In 1960, the United States sent Frank Chapot, William Steinkraus, and George Morris to the team competition in show jumping at the Games in Rome. They returned with a silver medal, and in 1968 William Steinkraus won the individual gold medal at Mexico City. These three riders changed the face of show jumping in this country, and thirty-five years after their team medal, they remain—as coaches, writers, and officials—the most influential figures in the sport. Something similar happened in combined training. After the advent of civilian teams, the United States won no distinction until Lana Du Pont, the first female rider in Olympic combined training competition, won the silver medal. Then in 1976 the team of Mike Plumb, Bruce Davidson, and Tad Coffin won the team gold. Plumb and Davidson, along with Denny Emerson, who came to the national team a little later, are still dominant voices in eventing. The difference between them and their older colleagues in show jumping is that they are still competing, and in this country, they are still the guys to beat.

In 1994, Plumb, who rode at least four horses in various divisions at the Millbrook trials, doesn't have a horse ready for the rigors of Fair Hill, and Denny Emerson's entry The Griffin has been scratched. But Bruce Davidson, the youngest of the trio at forty-three, has two running. Here also are David O'Connor, winner of Fair Hill the previous year, his wife, Karen Lende O'Connor, Phyllis Dawson, Wash Bishop, and Torrance Watkins, all of whom have represented the country on Olympic teams. There is Jane Sleeper, a diminutive person with a strong record and a remarkably light style, and an imposing phalanx of very tough younger riders—Stephen Bradley, Lauren Hart, Deanna Hines, Todd Trewin, Mark Weissbecker, Keith's friends Mimi Osborn and Missy Ransehousen, and Keith himself. The endurance phase, most likely the cross-country fences, will eliminate even some of the best of these.

The size of the fences is not usually itself a problem. When I walked the advanced course at Millbrook with Keith, we came to an enormous rustic bunker roofed with cornstalks. It was high and wide, and it was placed at the peak of a cone-shaped drumlin, so that the takeoff point was lower than the base of the obstacle. Keith looked it over but never stopped walking.

"The *size* of that thing doesn't worry you?"

"No, it's just big," he said, and in a few moments we reached the next obstacle, which he gave full scrutiny to, a set of stepped banks with a drop landing. He rehearsed the angle of approach coming from the big bunker on the drumlin, and he paced off the interior distances, then checked this measurement. The intervals between the banks did not work out for even strides, and the problem for the horse was to adjust to the spaces between the banks, to go long or to go short and still be poised for the next launch. Keith would revisit his strategy for this fence on his next trip around the course. It is customary for event riders to walk a cross-country course more than once, and it is not unusual, particularly on a course as tough as Fair Hill, for the competitors to spend hours rewalking the four or five miles before they meet the fences with their horses.

Keith has been anticipating the cross-country run at Fair Hill probably since before he sent in the entry check. While he is fully aware of the dangers here, he seems to look forward without intimidation to trying them. When I asked him several weeks ahead of the event if he worried about this fence or that fence on the Fair Hill course, he would say no politely without paying much attention to the questions.

I think the demon that stalks him is Play Me's physical frailty. Play Me hurt himself, making it impossible to continue normal training and impossible to know if the leg had become sound. Keith warned me not to try to visit him before his run on endurance day. "It's going to be pretty tense." Worry about the course? I wondered. "No, the vet checks," he said, and in the vet box before the start of the cross-country course, he was clearly anxious, impatient to get away from the veterinarians and grooms and run to the first fence.

The demands of combined training are so great that just completing a course without substantial penalties is evidence of competitiveness. For Keith, what was to be feared, what would constitute failure, was not being able to complete the course with a sound horse.

I worried about Keith's horse not being young enough

and strong enough or, worse, even breaking down, and in spite of the courage Keith and the other riders can draw on, there were two fences I didn't have the guts to even watch Keith attempt with Play Me Right. There was the Serpent, with its blind takeoff down into the ravine, its treacherous shale shelf where the horses landed, and the technical difficulties of the abrupt turn at the bottom of the gully. Maybe this was Steve Stephens's "goddamn cliff." I couldn't watch. The year before I had seen many horses fly over the panel at the edge of the gully and drop down into the creek bed. Maybe other horses and riders succeeded here, but I didn't know them personally. I would wait until Keith had come through the Serpent to watch other riders fly off the ledge.

Then there was the fence just ahead of the Serpent. The Springhouse was new that year. Like the Chesapeake Crossing, the obstacle much later on the course where Play Me Right would hang up on the log, it involved water. It was basically a small pond the horses jumped into and then had to jump up a bank and out over another log. Riders approaching this first water obstacle had two choices: they could gallop around the rim of the pond and dive over a relatively straightforward fence into the water, or they could take the short option, which could save a horse as much as fifteen seconds. But short was costly. The time-saving entrance to the water was an obstacle made of two gigantic trees laid on their sides and suspended nearly four feet above the ground. The trees were too far apart for any horse to clear them in one leap, but so close that the horse landing after clearing the first log had to take a tight bounce, even tighter than the bounce back on the hill at the pen, and jump even higher to clear the second tree and reach the water. The short path asked for one very precise move from the horse, just one, but it asked for this precision from a horse who has trotted eight miles, run a two-mile steeplechase course flat out, and now in this last phase of the endurance test has been galloping and jumping for four minutes. Someone who knew what a good horse can do and who had a mildly diabolical imagination had designed this water obstacle. It was the first water on the course. I was curious to know if any riders would try the short route or if they would all ignore it as essentially impossible. Early on cross-country day, a couple of hours before Keith was to start the distance sections of the endurance test, I waited beside the Springhouse. Several riders passed without even glancing at the tightly spaced logs. They skirted the rim of the pond and dropped in comfortably over the log on the opposite bank. Then came Bruce Davidson on a bay horse named Heyday. They were galloping very fast, and fifteen or twenty strides out from the big logs, Davidson sat up, simultaneously bringing the horse under him and turning. Heyday was galloping nearly in place, hocks pumping. The horse was popping off his front end until he was in the position Davidson wanted him: confronting the logs at the absolute perpendicular, and bouncing toward the two logs with powerfully compressed strides until on one of these bounces the horse cleared the first log. Heyday came down cleanly between the two, already beginning the next thrust of haunches, which launched him up over the next, higher log and into the water. The short option at the Springhouse could be done. After Davidson had come through without a wreck, I watched two more horses pop through it, but I dismissed the possibility that Keith would take Play Me—a smaller horse, an older horse with a newly healed injury—the direct route.

If you are going to ride, you have to ask. After all my years sitting on the backs of horses, this should have been apparent to me. But it was only after Lendon said so that I realized it had relevance to my riding, and in fact to how anyone gets on in the world. We were driving to Bedford to dinner, away from the indoor school where Lendon had spent nine hours exhorting students, demanding, pleading, threatening them if they didn't seek more from the animal under them. We came to a knotty little intersection where a steady stream of commuters rushed around the tight corner on their way to dinner. "You're going to have to be more aggressive," she said, "if we're ever going to eat." I realized the willingness to push forward, to push someone else forward, was lacking in my dealings with horses. Later when I watched Anne schooling students before their jumping rounds at Attitash, I recognized the same habit of thought. One young professional successfully negotiated with her horse a rail hung high above the ground. She and the horse had cleared the rail several times but without satisfying Anne. After this last effort, Anne said, "If you want to win, you've got to give yourself less room, ask for more at the end. Add a little."

If you are going to ride, you are going to fall. This was demonstrated to Keith at an early age, and he has a big, ragged scar on his upper arm as a souvenir. He also has a sneaky sense of humor. He told me about the accident that caused the scar. He was about five years old and riding his pony with a group of other kids and horses along a roadside. His pony spooked, threw him, and the rein caught around his arm. The rein had a covering of pimply rubber to keep it from sliding through his hands, and this nonskid rein gripped the skin on his arm, tightened, and tore into it as the pony dragged him along. I thought about the terror this had caused the little boy.

"Why did you ever get back up on a horse?"

"Probably my parents made me," he responded with a poker face. I reported this to his father, who has no interest in riding other than Keith's and who has endured many worrisome moments on account of Keith's passion, and he found the remark and my gullibility quite funny.

A fall at a fence is a fall at a fence, and whether the sport is combined training or it's show jumping, the fall can hurt you. I may have come to show jumping with the idea that it is somehow tamer than combined training, but I was soon set straight on that. Because of the garden-party decor of the arena, because the action is confined to an area smaller than a ball diamond, and because the horses and riders look so practiced in the motions of jumping, show jumping looks quite civilized, and it is less rough-and-ready than riding cross-country fences. But show jumpers win by taking the least time, the least possible room to maneuver, and the lowest, flattest clear path over the fences. The horses get very good at this, but they are always attempting something they haven't tried before. The course always contains new combinations of distances and turns, heights and widths. A slight misstep, a little giving way of a haunch, an imperceptible struggle between horse and rider, can cause a terrifying crash, entangling horse and rails and rider. Frank Chapot, who has ridden and raised some of the most successful horses in the sport, now coaches the U.S. show jumping team. I asked him if after all the jumping rounds he had seen in his life he still got a kick out of watching competition. He paused as if he couldn't believe I'd asked the question, and he said no, he did not in fact enjoy watching much anymore—his daughter was now jumping off against the professionals. Accidents, he reminded me, are not uncommon. They happen.

While it is truly rare for a rider to be killed in a crash at a fence, the severity of other possible injuries is cautionary: broken arms, legs, ribs, hips, and backs; punctured lungs; blows to the head; spine damage; brain damage; incapacitation. Once you've been hurt in a fall, the recovery involves more than the mending of bones and the healing of tissue. When I asked Anne what it is like to be hurt at a fence and to try to mount up and jump another big course, she spoke about the ethic of getting back in the saddle and the pressure riders put on themselves to keep riding. The fall may have done real damage. The horse's confidence, along with his ability to jump safely, may have been compromised. You may not be able to ride. You may not want to ride, or you may just be afraid that you don't want to ride. But if you do not ride again, how will you know who you are?

There is the expectation that a real rider will get back on the horse. The real rider will face up to the fall and will return to the saddle to ride down any fear. In the early 1980s, Anne took two falls in close succession, each of which resulted in a concussion. She looked able-bodied, and although she didn't feel right, she worried about missing shows, losing clients, and appearing to be a malingerer. When she tried to compete again, she realized something was still wrong, and after consulting a physician, took six months off. It was a probation. She was very young, not yet twenty-five, and she had already built a strong record for herself. While she was put out of action temporarily, it occurred to her that maybe she didn't need to go any further with show jumping, and she began a serious reconsideration of riding and direction in her life. But as she regained her strength, she began to look forward to riding, and by the time she was pronounced healed, she couldn't wait to try a big course. Time brought her riding back, and having learned that, Anne characteristically incorporates giving herself time and taking care of herself into her strategy for dealing with accidents. She also uses a helmet with a safety harness that secures it to her head. These newer helmets have become mandatory for any jumping in combined training and for the kids who ride in Pony Club, but many show jump riders resist them the way some motorcyclists shun helmets. Anne, however, has organized her responses to risk.

Every day in the long season of competition, Anne is balancing the chance of turning instantly short to a fence and clearing it against the chance of a crash. If there is one, she says, "my first concern is always the horse." Even though the fences will come down when a horse runs into them, a collision can cause the horse to fall. Depending on how extreme the loss of balance and impact for the horse and how game the animal is, the fall can rattle the horse's nerves, impair his confidence in himself and his rider. If Anne rides too fast and dangerously, she can leave the ring with a discouraged, worried horse or a horse with an outright injury. This risks the support of her owners and her future opportunities. She must do what she can to protect the value of the horse. But if she does not push the horse into the margin of risk, there is no possibility of succeeding. How far she ventures into riskiness depends mostly on how important it is for the horse to succeed in that class, on that day. Over the season I saw her ride in many classes like the one with Ferina, where the test was a builder of experience—and many where the horse was a good one and a contender over the course, where she took some calculated risks to come in fast. But it was not until the last outdoor show of the summer, the American Gold Cup, that there was enough at stake for her to take all the risks.

There aren't as many days on Play Me Right's calendar. Keith cannot spread risk over many competitions. Play Me runs five or six times over the season, and this year's schedule is crowned by the Fair Hill International three-day. Saturday, cross-country day, is the day when the risks are there to be taken, and foul weather or fine, good footing or no, the fences wait. "Once I decide how I'm going to ride a fence," he told me, "I can get pretty stubborn about it. I don't like to change my mind unless something big happens out there on the course."

"What he has on his mind is speed," Millie Taylor told me. "Ever since he went up to the USET clinic a couple of years ago, that's been what's on his mind. Speed. I really wish he would just let go of that and jump around the course." This is the sort of thing your mother probably should believe, and after enduring the anxiety of standing by obstacles for nearly every cross-country run Keith has made, she should be forgiven for not caring about time penalties. But Keith has to care. At the Springhouse he bounced Play Me Right through the big logs that were the short path into the pond.

We heard the whistles used to clear spectators from the course as another horse thundered down to try the pen and head off to the far regions of the course. Play Me was running in splendid, fast company. Horses in the early section of the course were jumping down through the pen on the hillside. At the same time, other horses galloped toward our hillside, toward the Chesapeake Crossing's dike and pond. Their riders gathered them up for the short run up the dike, and the horses hopped handily over the big log, down into the pond, up over the gate jutting out into the middle of the water. The second water obstacle wasn't so tough, it seemed. The second water could be done.

The whistles near us blew again and again to clear people away from the fences. Over the public-address system, the call of numbers and fences let us know that Play Me Right had cleared the first water obstacle, was on his way to the Serpent, and thirty or forty seconds later, was up out of the Serpent and running through the back of the course toward us. Another minute or so, and in the shadow of the pond dike, Keith is putting the whip to Play Me. He is asking.

Play Me responds. But his effort leaves him suspended momentarily on the log. He teeters back slightly. Keith uses the whip again, and the game horse pitches forward off the log in a little dive. Once he is down in the water, Keith gathers him up. They jump over the gate and gallop out of the water for the next fence.

"I thought it was over," I say as a group of us hurry

up to the crest of the hill to see Keith ride the sunken road and then the final fence. No one is listening. "I thought it was over."

At. the sunken road, it nearly is once more. This obstacle is really nothing to stop the catty Play Me. But it is very far out on course, and the horse is tired. Jumping out of the road, a front foot catches. It sticks for an instant, and the rhythm stops. Play Me's dark shoulder drops toward the ground. He manages to lurch forward, recover his stride, and leave the front shoe on the ground where it had been pried loose.

"I thought it was over," I told Keith later. "You got very lucky."

"You know," he corrected me, "that's just the way that horse is."

What They Will

Play Me Right struggled to clear the log and drop into the water, he picked himself up and kept galloping after his shoe was cast, and there are two explanations for this: Keith wanted him to, and he wanted to. Riding in the grand prix at the American Gold Cup, Anne needed the same degree of co-operation and determination from both Eros and Dynamite. Competition tests the wills of the riders and horses as much as it tests their physical capacities and skills.

Each of the three Olympic sports defines the importance of competition differently, and the divergence is sharpest between dressage and show jumping. I visited Lendon's stable one bright fall day when several students were riding outdoors in a large arena. A man circled the enclosure on a tractor outfitted with a front loader, and one of Lendon's grooms went behind the tractor with a rake, sorting out stones that were large enough to bruise the bottom of a horse's hoof. I found Lendon on her hands and knees in the middle of the horse traffic. She was after a cache of large stones left by the tractor. One of the knee patches on her breeches had worn away from the fabric and flapped at the top of her boot. The last time I had seen her, she wore snowy breeches, a formal black cutaway with tails, and a silk top hat. Now she was preoccupied by the stones in the sand underfoot, and she walked along with the young man using the rake, talking about her performance in the trials for the World Equestrian Games, talking about judges and judging, talking about competing, and squatting abruptly when she spied another land mine waiting to lame a horse.

Lendon Gray could have taken the princess track. She was raised in a middle-class family that was able to send her to prep school and then to Sweet Briar, where a good number of well-to-do and outright rich families sent their daughters. Even though the women's movement was beginning to change the expectations of young women, many of Lendon's peers were headed through college to big weddings and installation in handsome houses. If she had pursued those expectations, she could have easily had good horses and top-level coaching. Even now, the access many riders have to good horses is through the convenience of their sleeping arrangements. But I imagine that even in her college days it was difficult to picture Lendon shopping for lingerie or pondering thread counts. She is too Yankee, too rigorous, and too strong, and the force of her will is never more evident than when she is competing, when she is riding in front of judges. Nothing interferes, nothing distracts. She is, as she says, "nerveless."

"I am a good test rider," she told me. "I'm not that good a *rider*, but I am a good *test* rider."

Perhaps because competition comes easily to her, Lendon has come to a real ambivalence about it. Paradoxically, this has sometimes worked to her advantage when she is competing. In the spring of 1988, the USET would name its dressage team for the Olympics on the basis of two selection trials. "I wasn't sure I wanted to be on the team— I was nervous about going to Korea." But she went to the Gladstone, New Jersey, headquarters of the USET for two weekends. "We all rode two tests the first weekend. I brought my horse home. Everybody else stayed at Gladstone. They were training, training, watching each other's rides, training—and I was home trail riding." When Lendon returned for the second weekend of judging, "I rode the same two tests and scored about the same. Almost everybody who was ahead of me scored lower. I didn't ride any better. They just didn't ride as well. So I ended up going to Korea."

They beat themselves.

After years of riding before international judges, Lendon sees competition as something apart from really riding, a way to test the validity of how well you are riding—and riding is what is most important to her. I once asked her if, dealing with as many horses as she does, she develops strong emotional attachments to a horse. Not usually, not anymore, she told me. "It hurts too much. You put your heart and soul into a horse, and then they take the horse away from you. Now I put my heart and soul into my riding." This is a life work. As she sees it, horses—*any* horses—will come into her hands, and they will get better. She will get better. It is a process without a finish line because dressage has a cruel standard, perfection. Along the way, one of these horses may be of Olympic caliber, but she says, "I don't know whether I'll make another team—and right now, I don't know that it matters."

Unlike dressage, where the ideals hold equally whether the discipline is practiced in competition or whether it is simply practiced, show jumping is entirely dependent on competition as its standard. For Anne Kursinski riding on another Olympic team is the single, unqualified goal of her work now. Immediately after the Barcelona Games, where she was disappointed, she began to engineer a trajectory to take her not only to the Olympics but to a position of real competitiveness there.

The year I visited her, Anne came off the winter season in Florida, where she had been injured and kept out of competition, with too few points to make the list to train with the USET in Europe. But late in the spring, her win at Devon with Dynamite put her on the list. She went back and forth to Europe twice, flew to Calgary for an international event, and kept up her domestic competition schedule. She would get off a plane, drive to meet Market Street wherever the show was, and go to work with the horses and with the numerous professionals and amateurs she coaches. She was on a horse by seven in the morning, exercising—then warming up, coaching warm-up, riding courses, critiquing students' rides, meeting clients, checking in with the grooms, and, finally, tending to some detail of a horse in a temporary stall before finding her motel room. I suspected that at times the fatigue was stunning— I imagine her on an airplane closing her eyes as the plane taxis away from the terminal and opening them when it bumps down on another continent.

If I was a CEO and Anne Kursinski was hired into my organization, I would dust off my résumé and have lunch with a headhunter. Anne would get my job, and I would have to feel okay about it, because it would be the best thing for the company. She is bright and completely clear about where she is going. She long ago accepted the fact that she needs other people to help her get there, and she values their contributions. I find it telling, for instance, that she refers to the grooms at Market Street as professionals. Systematic attention to detail, relentless drive, unblinking honesty about her own performance, are packaged in a steady, sunny person. They carry along her organization—Hoffy, Rob Spatz, Carlos Covarrubias, and Stacy Falco, whose working lives are devoted to the painstaking

care of the horses assigned to them. Market Street is essentially a company. It was formed and developed to allow
Anne to compete, and it is dedicated to winning.

What sustains Anne's personal efforts and fuels Market
Street may be will, but she understands—as do Lendon and
Keith—that human will is only part of what is necessary.
There is another will involved—the horse's.

Anne Kursinski may want more than anything in the
world to ride in the Olympics, to win the Olympics. She
may tax her managerial skills to their limits and drive herself physically until she is past even exhaustion. But she
cannot get as far as the first qualifying class if the horse is
not willing. The talent to fly over fences belongs to Eros.
In fact, Anne told me, "I never thought of myself as an
athlete until after the silver medal in the '88 Olympics. I
always thought of the horse as the athlete."

"Very willing" is a phrase that comes up often in copy
for ads offering horses for sale, and it is true that some
horses are more willing than others. They are more cooperative. They may have fire, they may have quirks that
make them difficult, but eventually they will give greater
effort to their riders. They will go to work happily. This is
essential in a good horse. Lendon told me about a talented
horse she trained up through the highest levels of dressage.
She competed him successfully for a season, but as soon as
this was accomplished, she sold him. "He wasn't happy,
and it was just such hard work. It was no fun."

Willingness has been a preoccupation of the masters since Xenophon, who instructed riders that "when a horse suspects some object and is unwilling to approach it, you must explain to him that it is not terrible, especially to a courageous horse." They have resorted to some imaginative means to counteract nappiness, the essential unwillingness to go forward: a cord tied to the horse's "stones" or testicles, for instance. One sixteenth-century Neapolitan theorist prescribed a live hedgehog tied by its leg under the horse's tail. But since the mid-eighteenth century, it has been generally accepted that horse happiness is the key to cooperation.

A happy horse is more willing, and each of the trainers I visited went to great lengths to make sure their horses were comfortable and content. They monitor such details as the horse's time at pasture, which horse is stabled next to which, and how much satisfaction the horse takes from his food—Hoffy once excused herself from conversation to take care of an important piece of business for Market Street: "Fifty pounds of carrots," she said with a conspiratorial grin.

In Florida, Lendon would be wakened every night by Medallion kicking at his stall. He let fly with a regular, steady tattoo, and Lendon had to keep boots on his hind legs to protect him from his own displeasure. He had not been castrated yet and had even fathered some foals. It isn't at all unusual for a stallion to have a dual career as a competition and breeding horse, but "he wasn't happy as a stallion," Lendon said. Eventually she arranged to have him gelded, not to subdue him but to put him at ease.

It is often said that geldings are the most willing gender and stallions are more cooperative than mares—or that a mare's hormonal activity makes her downright difficult —but neither Lendon nor Anne held with this. They thought that while it might be more difficult to find the key to a mare and while stallions might tend to hold back some increment of effort and not give their all, a horse's willingness was mostly a matter of the character of that particular horse.

Sometimes horse happiness depends on what job the horse is assigned, and career changes are common. Owners become frustrated by their horse's lack of progress and decide perhaps the animal would be happier in another line of work—a jumping horse is sent off to study dressage, a polo pony becomes an event prospect. But other times, a change of duty is just a superficial fix that doesn't alleviate a deeper unhappiness.

Poteen was a rawboned, Roman-nosed Irish horse with bright markings. At the age of fifteen he was donated to the Equine Research Park at Cornell University. Retirement as a research subject was the final resort of his frustrated owners, who were the last in a string of frustrated owners. Poteen had fallen to this anonymous place from a high-level career in international show jumping. His work had been supervised by the best trainers in the United

States. But Poteen (the Irish word for moonshine), like whiskey, held the capacity for charm and violence.

When he arrived at the research facility, he betrayed neither. He was dignified, remote, and very capable. The staff knew that Poteen's trouble had come with jumping, and they offered him to my friend Debbie with that warning. She wanted a horse for dressage, and Poteen had the size and power. She started with him by going back to the beginning, as if he had never had a day of training. He was advancing well, beginning to progress through the movements of the lower-level dressage tests, when Debbie mentioned his name to a well-known coach, who remembered the horse.

"Poteen?" he asked cautiously. "How are you getting along with him?" She recounted their accomplishments, and the coach said encouraging things. About the earlier Poteen he said, "He didn't like his job. He hated the work."

Debbie had changed the horse's job, and this seemed reason to think she might be successful with him. She and Poteen tried more difficult movements, and once during an indoor coaching session her instructor stood up suddenly and Poteen whirled away from the wall. He wrenched Debbie off his back and sent her hard into the ground. Her pelvis was broken. But he showed no signs of repeating the behavior. So she and her coach decided the incident had been an accident: Poteen had been frightened. Their work with the horse went on. Debbie began to compete Poteen, to ask for more intricate responses. She took the horse to a prestigious competition in Saratoga, New York, and in the warm-up area there, he reared and whirled and plunged like a mad thing. The fall she took broke her back. Poteen had reached a sticking point. He had reached this point often in his troubled life. He was unwilling, and this was a state of hysteria in which he could not progress and could not tolerate even the expectation of progress. Debbie could not identify precisely what triggered the tantrum, and rather than return him to end his life as a research subject or put a new owner at risk, she had the horse euthanized.

Poteen's story is a dramatic enactment of a limitation that is ordinary to the horse. There are horses all over the country who have been turned out to pasture because they don't *want to*. They don't have the will to try what they're asked, or perhaps their riders aren't clever enough to keep them happy and bring out their willingness. These horses are ordinary, just as people who haven't the motive to try for something beyond easy reach are ordinary. This is not to lay judgment on either the ordinary horse or the ordinary person; it is just to point out that a freely willing horse is relatively rare. Usually this kind of horse is one that has been brought along by a rider with insight and the will to act on it.

This is certainly the case with Lendon and Last Scene, a horse she describes as both "feisty" and "generous," and whose first years in training were long and exasperating. It is also the case with Keith and Play Me Right, a horse that was unhappy and difficult about jumping until Keith figured out how to make him happy about what he did so well. With her two more seasoned Olympic prospects, Anne has had two contrasting personalities to placate, and she has had to learn how to keep Eros and Dynamite happy even when the pressure builds because she really wants the win, the way she did at the Gold Cup. Her rides there were a real-time diptych of horse wills moving to human will.

Like many Thoroughbreds, Eros is wired, and he is an independent. He was purchased by a group formed shortly after the Barcelona Games to buy a horse that could carry Anne to the Atlanta Games, and he is the horse in whom her Olympic dream is most vested. By the time Anne discovered Eros, a number of other strong international riders had considered him without making an offer. His form over jumps was hardly classic, and he didn't seem to have the malleability that allows a show jumper to turn and bend and spring in fluid synchronization with the rider. But he appeared to know his job and to have his own concept of what constituted being good at it. He arrived at Market Street and lurked in his stall with an evil expression. Anne's response to this was to give him space, to give him credit for what he knew, and since he was standoffish, to leave him alone.

When I saw Eros, he had been with Anne two and a half years. He was somewhat more engaged by activities in the stable, but "still aloof," Anne said with a smile. She had learned how to shape his routines to keep him happier, and she had developed an even greater respect for his ability to figure out tough jumping problems and for his confidence in his own skills. Over the summer Eros demonstrated to Anne that he was Olympic material. In Rotterdam, he jumped some huge fences to finish eighth. Anne said the fences might have given her pause when she walked the course, but that it doesn't occur to Eros that a height and distance might not be possible. If there is a fence in front of him, "he will try."

She described Dynamite as less indomitable than Eros, more easily distracted. He seemed to need more support from her, and for this reason, she had less hope invested in his Olympic prospects. He was regularly beating the best horses in the United States and had won a number of big grand prix. He had held his own in Europe. Even so, she was trying to be realistic about the demands and the chanciness of the long qualification process in 1996. There would be eight qualifying trials over a nine-week period in the spring. So many rails, so many clocks to beat, and she knew that even if a horse survived the qualifying trials, he would still face the size and technicality of the courses built in Atlanta and the strength of the European horses.

The Gold Cup came late in a hot droughty summer, when nearly all the shows suffered from dust and from hard footing made slick by grass that was dry as excelsior. The show is staged at the quaint fairgrounds at Devon, just south of Philadelphia. They are now closed in on every side by suburban homes. The main grandstand, the pavilions, and the settlement of tiny shops are old but freshly painted wood structures. There is a cupola on every roof that will accommodate one, and except for the newer auxiliary grandstand and stables, everything is white with baby-blue trim. This snug setting hosts some of the biggest show jumping audiences in the country. At the center of the grounds is an oblong arena, the Dixon Oval, where Dynamite had won big in June and Eros had won the Gold Cup the year before. This year California course designer Robert Ridland has banked the Dixon Oval with voluminously massed chrysanthemums, yellow-gold and deep rust, fat marigolds, and lavish splashes of foliage, silver-laced leaves and magenta-streaked spears. The fence standards rise from similar banks of bloom and colored leaves, as if they grew there. I cannot begin to find a number for the flower stems or even for the pots.

The boxes in the old grandstand are tiny, but people squeeze happily into the narrow wood folding chairs there. A wide plank on the front rail of each box serves as a table, and Friday night there are picnic baskets, hors d'oeuvres plates, and wine bottles balanced on these planks. There are many locals in the crowd, many people who have a horse or two at home, and many who come because they host friends here every year. As it grows dark, the lights in the arena begin to seem quite strong, and the hors d'oeuvres begin to be supplemented by chicken barbeque, sausage and pepper sandwiches, and funnel cake, an elaborate nest of deep-fried batter coils drifted with powdered sugar. The qualifying class for the grand prix begins, and people continue box-hopping until the jumping is nearly over.

A week earlier Eros had competed over a challenging international-level course at Spruce Meadows in Calgary, and Anne was concerned about how much the jumping and the air travel had taken out of the horse. To give him an extra day of rest, she asked Rob to leave Eros in New Jersey until Thursday afternoon, then go back with the van to bring him to Devon. When Eros trotted into the lights Friday night, he betrayed no fatigue. Under the lights, which show the horses and fences almost as clearly as daylight does, his copper coat glimmered. The lights throw down shadows and put the audience in a dark recess, but they did not bother him. The horse radiates energy and keenness.

There were about fifty horses entered, and their riders included most of those contending for places on the team that goes to the Atlanta Games. After the course walk, I looked over at the in-gate and the VIP seating at the other end of the oval and realized how much of the United States' expertise in show jumping was concentrated on this old fairgrounds. Sally Ike, the USET director for show jumping, is on the judge's panel. Two Olympic veterans who will coach the team that goes to Atlanta are here. Frank Chapot is the chief judge, and George Morris is coaching several riders. Another Olympic veteran, Mary Mairs Chapot, has been walking the course with her daughter Laura, and among the other riders there are at least fifteen who have represented the country on USET teams. Of the ten riders considered most likely to make the team for the Atlanta Games, only Hap Hansen and Eric Hasbrouck are absent.

To the list of human experts, I should add the name of one horse, Gem Twist. Raised by the Chapot family but competed by others until now, this big gray gelding won silvers in both team and individual competition in Seoul and he has been named Horse of the Year by the American Grandprix Association three times since 1987. He is sixteen now and has just won another grand prix with Laura Chapot riding. She is dark-eyed, quite thin, and although she is in her twenties, she looks much younger. Still a rookie, she is one of the most tactful riders I've seen and over the summer she has begun to use her remarkable lightness to strategic advantage. But the crowds don't care who rides Gem Twist or how talented she is. They adore the horse. They revere him more than dressage fans revere Gifted. His presence here—or at any show—makes it seem like big-time jumping.

The qualifying class was not stringent in the cut it made. Thirty-five of the fifty horses would succeed to the grand prix. Eros came early in the lineup. He jumped around within the time allowed and left all the fences up. His time was one of the fastest in the first round, but it was bettered a few minutes later by Dynamite. The two of them looked good, but Anne did not really push them in the jump-off rounds. She was conserving horse power. The qualifier, itself worth $25,000, was won by Peter Leone, a rider who recently rejoined the professional ranks after trying a career in the stock market. His horse, Legato, is a strapping bay gelding who, like Ferina, is a model of equine beauty and has become a threat to the horses established at the top.

On the morning of the grand prix competition, we had the first real rain of the summer, and it was a big one. A heavy downpour, the kind that comes in off the ocean when there are tropical storms in the south, fell on Devon. The fairgrounds were awash, and except for the crew in hooded rain gear setting rails in the jumping arena, the spectator areas were deserted. The rain was too heavy to let me recognize the single horse and rider splashing across the warm-up arena. I went around to the old barns where the Market Street horses were stabled. Anne had been there earlier and exercised Eros and Dynamite in the downpour.

With his head reaching to the middle of the dim, narrow aisle that ran between the stalls, Eros stood to be groomed in the light of a single bulb. Three stalls away under another bulb stood Dynamite being made immaculate by Carlos. Both horses are bright chestnuts. Eros is slightly redder, and Dynamite is slightly heavier in build. Dynamite stood quietly, but Eros wriggled under Rob's curry, never still. Not much dander flew up from the edge of the curry, and this explained something about Anne that had caused me to wonder: her fingernails are always clean. It is very difficult to handle horses, even horses that are groomed and bathed regularly, and come away without dull gray debris under your nails. These were remarkably dirt-free horses. Eros stepped about restlessly, moving to the limits of his cross-ties, pressing into his halter. He was very active with his head and passed his muzzle fleetingly at Rob's shoulder. "He won't really bite," Rob said. Eros likes play. He likes to touch Rob and he likes Rob to touch him.

Dynamite has a bold bright stripe down his face. "And he *will* bite," Rob told me. Carlos laughed and nodded. Dynamite wore a look of absolute innocence. Evidently he waits for his opportunities.

"Do you think she'll ride?" I asked Rob, concerned about the downpour outside.

"It's supposed to stop this afternoon," he said, but still I wondered if she would risk the horses. "There is that," he agreed about the slimy footing, but to him, scratching didn't seem a strong possibility. "She'll walk the course," he promised, and it would be more challenging than the course for the qualifier. The fences would be taller, the distances trickier, the turns tighter.

The rain blew off just before the course walk. It left water standing in little pools around the arena. But water doesn't change the procedure. The arena is well drained and the material in the footing the safest under wet conditions. It is finely granulated stone and doesn't form mud or ooze when mixed with water. Still, there was a lot of water in and on top of this footing, and it would be slick going. The course was crowded with riders and coaches in damp clothing and waterproof footgear. The entry was a full complement of the thirty-five horses who had qualified in the warm-up competition, and because of the importance of the Gold Cup, the competing riders were joined in the course walk by some professionals who had no entries but wanted a look at the course. It looked big to me, twelve obstacles at full height, some deceptive spaces between them, some tricky internal distances, and the customary set of switchback turns. The riders spent a long time walking through the progression of fences, then breaking the course' down into its components. They measured and analyzed these: the rails hanging between replicas of the Liberty Bell, the two Shamu killer whale figurines guarding the fence over an artificial aqua pool, the brick walls, then rail after rail of the double and triple combinations, and, last, the big red Budweiser panel suspended between gigantic replicas of long-neck beer bottles. The riders were striding off, counting, re-counting.

Walking a course with Anne while she is coaching students is elucidating. You learn how she would plan distances for each particular horse, which fences she considers potential upsetters, and which she would gallop over, business as usual. But when she is walking for her own rides, she turns up the imposing force of her concentration and willfully shuts out thoughts on any topic other than jumping these particular fences. She paces off the striding and counts. She talks to herself—later, in the press box, they tease her about whatever it is she utters as she rides into the ring, apparently the numbers she has told herself during the walk. She pays no attention to other riders, other people. If she doesn't know you well, she will probably fail to recognize you. It is as if she has removed herself to a space where she is allowed just to observe and consider a particular situation in all its aspects. I've seen this remove even early on days when there is something important to be done. She has gone off somewhere to prepare.

Anne was working out strategy, checking her impressions of distances, talking to herself, long after Hoffy had dropped away to chat and work on a sandwich. Later, after this group of riders dispersed to go back to the stables, Anne and two other competitors returned to reexamine the course. One of these riders withdrew his two horses, presumably because of the footing, and within a few minutes three more scratches were announced. But in spite of the risk, most of the field remained in competition.

They would be allowed eighty-one seconds in the opening round, and those horses that left all the rails up would return to go for time over just six of the fences. Eros came late in the lineup, followed still later by Dynamite, and before Eros trotted into the ring, five earlier horses had conquered the course with flawless rounds. Eleven had failed, including Gem Twist. Most of the rest of these "failures" were top horses that had been in the money frequently all summer.

The jumping horse has to be fast and perfect for a given minute and some seconds. Sometimes, his performance depends on which minute and some seconds he is given. Extreme, a mare that had been competing successfully against the Europeans for Anne's friend Leslie Burr, knocked a rail out of its brackets, and that was it. She was out of the competition. For the elegant Legato and Peter Leone, who had won the warm-up competition, the ending came as a triple blow. Three fences down, and he was done. I had seen Eros when he was not "on," and I wondered about the burden of fatigue the intensely competitive summer had put on Anne and both horses.

If either the horse or the woman was tired, there was no evidence of it. In fact, Eros appeared to have made his own strategy about distances and strides and to be intensely conscious of snapping his legs tight to his body. When he came straight through the triple, Anne stayed with him quietly, and I could see what she meant about leaving him very much to his own devices, letting him make many of the decisions. "You know, I don't really ride him. I don't mean I let him run around the jumps on a completely loose rein, but I don't *do* that much. We just kind of figure it out together—there's something Zen about it."

Nothing fell. Eros's time was respectable. He could come back and go for speed.

Dynamite's trip around the fences was smoother. His jump is always a rounder movement, and he made less commotion in the tricky distances. Nothing fell, and Dynamite was a little faster even than Eros. Of the eight horses that would jump off for time, two would be ridden by Anne Kursinski. Each of her horses had won a grand prix in that same arena during the past year, and now each would have a chance to put the six jump-off fences behind him as fast as he could.

The first couple of horses to race around the jump-off course post times of forty seconds or above. One horse topples a rail. Then Margie Goldstein, a popular rider from Florida and a top money winner, breaks the forty-second barrier. Immediately after her round, Flirtatious, a tiny mare ridden by Elizabeth Solter who jumps as if she's working off a set of springs, cuts nine-tenths of a second off Goldstein's time: 38.87 seconds, the time to beat.

Eros comes back to jump off, and independent as he is, he bends to the Kursinski outline. As he trots in, Anne is speaking to herself again, working on a genteel windup for the first fence. She is riding quietly. But once Eros meets that first fence he is tearing. She is still quiet but quite low and tight to him. Zen controls. The corners are trimmed, the distances shortened. Eros scrambles joyfully through the double, head up and looking for the next one, then the last one, the beer bottles: 37.795 seconds.

Eros has it won. Anne's pleasure and his pleasure in her pleasure are evident. This will make him a two-time winner of the Gold Cup. Even though the time is there to be bettered and there is one more horse to jump off—Dynamite—the spectators call it a win, and the television commentator calls it that way too. Then Anne does something I find astonishing.

Eros is partly owned by Anne Kursinski, as a member of the Eros Group. Hoffy is another member. Their horse is now, for all practical purposes, a two-time winner of the Gold Cup, a $50,000 grand prix. Dynamite belongs wholly to a longtime client of Anne's, Alan Shore, Jr. He has already won plenty of money for Shore in 1995.

Her horse's time is there to be brought down.

Anne reappears in the oval arena with Dynamite, and it is clear from the instant the horse triggers the electronic eye to start the clock that she is riding to bring down the time. She rides Dynamite quite differently from the way she rode Eros. He is more tentative, and she pushes him. She uses her legs and her hands. She sends Dynamite on with her voice. She wants that time to come down. She snarls. She rides like a cowboy. They take the risky inside turn she took with Eros a slice faster. She wants that time to come down. When Dynamite passes the electronic eye at the finish the clock freezes: 36.39 seconds. She beats herself. Carlos, not Rob, will take home the groom's award, a curly maple tool cabinet.

Dynamite is as good as he can be—on that day, in that half minute. Who knows how good he can be, how good Eros can be? Even while she's in the air over the fence, a show jump rider has to turn her eyes toward the next fence, and that afternoon at a press conference Anne is asked what she thinks of the course. She must be thinking about the ride she has just completed in terms of preparing for the tests ahead for her horses, the demanding series of selection trials and the Games themselves. She pays tactful compliments to the course designer and says, "I thought the oxers could have been squarer." Raise the rails. Increase the distance we must go. We will. They will.

Up-Downers

In 1995 Lendon Gray bought a house. It is the first property she has owned. After more than twenty years of moving stables and living quarters, gaining students, losing students, piecing together funds for international competition, she has reached a kind of stability, and now, although she is still approached by people who have a colt for sale and people who would like her to buy team jackets with "Gleneden" embroidered on the back, she is also contacted by brokers selling investment strategies. She was filling out a form for one of these concerns—what is your net annual income? assets? long-term liabilities?—when she came to a question she couldn't answer: at what age to you plan to retire? *Retire?* Stop riding? Stop teaching? She couldn't begin to apply those ideas to herself.

In this, Lendon has lots of company among horse

trainers. They do not plan to stop what they're doing. Partly this is because a livelihood from horses is usually tenuous, and trainers are rarely in a position to plan for the long term. Only recently have they begun to recognize that it may not be necessary to die with their boots on.

But there is another reason Lendon and many others do not plan to stop: training is seductive. It is a process that has no end point and that is always just beginning. How can you stop just now, when the young horse you have taken on has such enormous potential or while your most dedicated student has reached a critical point? There is always more to learn or, as Lendon points out, to relearn.

For the humans and the horses, riding and training are a continuum of innocence and experience. When you first sit a horse, you are aware of the precariousness of your position. You lose the power to ambulate, and you have the sense of hanging from your crotch and teetering there, the fulcrum of an uncertain pendulum. But the horse seems not to notice your weight or your will. His body is warm, even on the inside of your leg where the saddle flap intervenes, and you can smell him. It *is* precarious. You are told to squeeze your legs against his sides. The contact is reassuring, but when the horse moves forward, you are likely to be as astonished as the horse who first receives a rider. You look down and find an odd perspective. The ground is moving far below your feet. You look up, and you find the remarkable view ahead between the horse's ears. This is the horse's perspective, and after you have mounted up several times, you begin to accept that the animal is a solid connection to the ground. You expect to take that view.

You will do this over and over, and you will be a beginner. If you are lucky, a school horse will carry you through these first experiences. School horses are for the most part—or until they burn out as if they were social workers—stolid types willing to tolerate unintentional physical assaults on their backs and sides and mouths. There are legions of horses to fill these positions. It is honorable duty. They walk in drowsy circles until their passengers achieve some degree of balance. They jog lazily through the more uncomfortable phase during which the rider learns to rise in rhythm to the trot, popping mechanically out of the saddle, then looking down at the horse's shoulders. A school horse's job is to endure while you gradually order the activities that make it possible to actually ride. At this point, when you can rise in rhythm to the trot, you are what Keith calls an up-downer.

If moving in time with the horse appeals, you may continue until you learn to settle into the horse's canter stride, then begin with fences the way a learner horse does, by trotting over a row of poles on the ground. You will be shown how to hold yourself up over the horse's shoulders. You may go on and learn to ride over courses or you may find yourself content to just ride out in the open. But at some point in this process, responsibility for what happens between you and the horse will begin to shift. You gradually take the reins.

For some riders, this is the most difficult transition in riding. At Lendon's arena it is not unusual to see a rider with just the basic skills on a green horse. This goes somewhat against the received wisdom that pairing an inexperienced horse with an inexperienced rider is risky business or, at very best, the slowest possible way for the rider to learn. But Lendon says this rider is often better for the horse than a rider with enough education to try to force high expectations on the inexperienced horse. She doesn't appear to worry about the welfare of either the up-downer who pops up and down beside the more highly skilled riders or the green-broke horse. She gives this pair more time unobserved, which is time to experiment and find out how their movements affect each other. All the up-downers have to do is mind the traffic patterns in the ring and give way when custom dictates. Occasionally, Lendon will call out some simple command—"Left rein . . . left, *left!*"— and then return to the students who are attempting more complex moves.

At noontime on a very warm late spring day, one of Lendon's working students brings a horse to the mounting block, steps up, and starts out into the arena, where one other student is working. There is an apple tree near the arena fence, and under it a picnic table and the morning crowd, the women who have already finished their lessons. They are eating bag lunches, sharing chips, and happily passing the time. Fifi Clark is among them. She nearly always is. Every day except when family obligations call her elsewhere, she is at Gleneden riding her big, heavy-going mare, except that now, two years since the first day I'd watched her ride, the mare isn't so heavy-going. Fifi is riding her and competing successfully at the upper levels of dressage.

Neither the working student nor the horse has much education. She can rise and fall to the trot, but that is her limit. The reins flop, her legs flap. No part of her body has secure contact with the horse. For this pair, turning across the corners of the arena is like coming about in a sailboat. Her bouncing to the trot is high and uncontrolled. In one of her descents, she jolts down into the saddle and the jarring there causes her hands to jerk upward and snatch at the bit in the horse's mouth. The horse's head jerks up just as the woman is rising again, and it smacks into her nose and mouth like a prizefighter's punch. The student gasps. Lendon turns to check this out, and in the student's next breath comes rasp, a wail. It continues sounding even as the horse bumps down into a shambling walk. Several of the picnickers rush to the arena.

"Leave her alone," Lendon orders, and the rescuers stop at the rail. "She's all right." The wail continues for a couple more seconds, until the colt brings the student to where the women are looking anxiously on. "I am going to fucking *kill* this horse," she announces, and the women who would have helped her freeze. It is not her intent but her language that is shocking. Even though you can go to the mall and hear a full complement of foul language, you do not hear it at Gleneden. Lendon herself appears not to hear it. Her student approaches on the horse.

"Lendon, what should I do?"

"Cool off," Lendon suggests. "Don't curse your horse because *you* can't keep him going. Make him trot."

After a long ten seconds, the horse begins his loose-legged trot again, drawing a wavering line down the opposite side of the arena. The student looks back for help.

"Lendon," she cries, "what should I *do?*"

"Cool off," Lendon says with more emphasis, "and make the horse go forward."

If she is able to do this, the student will begin to know what it is to ride, and she will begin to have the skill to give this knowledge to a horse.

The education of the horse begins more gradually than the initiation of the rider. In ideal circumstances, humans will handle a foal as soon as it is born, and they will continue working through simple routines like leading in hand, lunging and driving on foot with long lines, that establish expectations and authority. Then when the horse is strong enough comes a telling moment in his education: he is backed. "Are you familiar with this term 'backing'?" Keith asked me. It is British usage, more respectful of the horse than "breaking." The horse's first experience of a rider is a moment of initiation. It is the beginning of a process that will continue over a long period with the introduction of more experiences that are potentially shocking to him. When he has accepted this other animal straddling him, he will begin to go forward. Often he is uncertain, sometimes fearful. He wavers. The path he makes wriggles. Sometimes he scoots or bolts. Further experience settles him, straightens him. When he is essentially rideable but not yet accomplished, he will be "green broke." Then he will be "green to fences," and just before he is fully trustworthy, "still a little green."

This is the stage in a horse's life that Lendon most enjoys. "If I could make a living at it, I'd ride nothing but four- and five-year-olds," she told me. As it is, she rides more greenies than most established trainers, and I suspect that the rapid rate at which the inexperienced horses learn offsets some of the frustrations of trying to help human students advance. Once she told me to stick around a while if I wanted to see a really great young horse, and late in the afternoon she brought out Jester, a four-year-old just purchased by one of her students. There was nothing particularly spectacular about Jester as he stood there. He was a sturdy horse with an only faintly decorated face. But once he began moving, Jester was on his way to being beautiful. He had been taught the basics, and for as long as he could keep his mind on the business of carrying a rider, his straightforward efforts brought real style to his movements. Any sight or sound could bring him instantaneously to a dead stop. When two people standing near the barn began to converse, Jester jolted Lendon nearly out of the saddle.

He didn't shy or bolt. He wasn't afraid. He just stopped to
figure out what he was hearing. "It's the walk-and-chew-
gum problem," Lendon said, and she found this hilarious.
She continued working with him, occasionally coming
loose when Jester stopped to take a good look at something. "It's rare that I get on a horse and I just don't want
to get off," she said quite literally in passing to everyone
who had now gathered to see the young prospect. Jester
stopped and lurched, and in between he went ahead, pow-
ering off all four. Lendon was unseated several times.
When she was done, she threw down the reins, lifted her
arms like a diva acknowledging applause, and demanded,
"Isn't he wonderful?"

It will be a good long while before most people can
see that Jester is wonderful. He is being educated for dressage, which can take years if it's done correctly, not "put
'im together so ya can sell 'im," as Lendon describes a more
hasty, superficial training process. In fact, Last Scene has
been working with Lendon for ten years now. The little
gray seems so self-assured and dynamic, but to my frustration, Lendon, having taken him to selection trials for the
World Equestrian Games, didn't seem much interested now
in competing him against her peers among FEI riders.
When I asked her about this, she agreed that Last Scene
is developing to his full potential. But she pointed out that
he would be a small brilliant horse going up against large
brilliant horses, that the effort of this kind of competition
could take a lot out of Last Scene, and that he was Peggy

Whitehurst's last horse. Lendon wants him to keep getting better and better but to last as long as he can.

The making of a show jumper is not so arduous. If the horse has talent, his education comes together relatively quickly, and if he is cared for conservatively, his competitive career is likely to last longer than the dressage or the event horse. His training develops through competition and progresses along a spectrum of greenness. George Morris has commented that green horses contributed to the disappointing performance of the American team at the Barcelona Games. The American horses lacked the necessary experience in competition. They were green to Olympic-level courses. In that sense, Eros is still a little green, Dynamite greener, and Ferina is quite green. Anne's trips to Europe and Canada were intended not only to allow her to become more familiar with world-class courses and to size up the European riders she might ride against in the Atlanta Games, but to give the horses that experience, to expose them to the only kinds of challenges that could take the greenness off.

Frank Chapot pointed out to me that Olympic courses have changed a good deal since he was riding over them. The design strategy used to be to simply present a series of enormous fences, but for the brave horses who succeeded over these courses, the burnout rate was high. After the Games, the competitive lives of these horses were apt to be limited. The design of the course for the Atlanta Games will be more sophisticated in its concept, but it will still be the toughest kind of test. The fences will be big, and their layout will ask for consummate skill on the part of the horses. By the time the selection trials are over in June, Eros will likely be the horse that can best face a test of this sort with Anne. But she doesn't rule out the prospects for Dynamite or even for Ferina, who is still a novice in the big time. It is possible, Anne thinks, that Ferina could make the most of her experiences over the winter and during the selection trials. She is backing up her chances on Eros with two other talents and giving them all as much experience as she and they can absorb without duress.

Although Anne's final preparations have consumed her for the past two years, this is still a relatively short phase in terms of horse education. To bring an event horse to the world class usually takes longer. After Play Me Right's earlier owners put in the time necessary to work the horse up from Novice to Training to Preliminary, Keith spent three more years bringing Play Me Right to the higher levels of the sport.

Now Play Me Right has gone beyond experience to age. He is nineteen. His performance at Fair Hill had been generous, and Keith decided to retire him. Faktor had begun to overtake Play Me. He was competing successfully at Preliminary and in the year ahead would move up to Intermediate. Although progress in combined training is never assured—"That's the great thing about this sport," Millie Taylor says, "it takes you so long to find out what you can't do"—if Faktor continued to learn and apply himself with the same willingness he was showing now, the youth and strength of the big red horse would make Play Me redundant.

Play Me hung around for a few months, and then one of Keith's clients observed that it was a shame the horse was not allowed to do cross-country anymore. He loved it so much. Why not run him at a lower level? She offered to foot the competition bills, and Keith brought Play Me out of his momentary retirement. They took the horse to the Loudoun horse trials, and at the first sight of the cross-country course, Play Me Right went wild. He hopped and plunged so much that Keith had to ask for help to get him into the start box. He smiled about the joy this brings Play Me, who now looks more like an old horse than he did a year earlier. There is a hint of pottiness about the belly, and his eye has a cast that is typical of age. The spark is not as bright. "I suppose it's a risk," Keith said about letting the old horse race over cross-country fences, and he thumped the dark neck affectionately. "So what am I going to do with you, hunh?" It is clear that Play Me Right does not understand quitting. "What do you think I should do?" Keith is joking, but the question is real, and it comes up for all competition horses.

"You can't just turn these horses out to pasture," Lendon says about the top-level dressage horses, who have been educated through a long, intensely personal relationship with their riders. "These horses are *people* horses." Her answer—actually dressage's answer—is the "schoolmaster."

The horse resigns from active training and becomes a teacher himself. When Medallion had achieved what he could for himself, he began to compete with Lendon's students, showing them his best grand prix moves, and now, working out stiffness every time he is ridden, he has begun to initiate the passive Karen into the higher realms of movement. Lendon gets impatient with Karen because she doesn't ride him assertively, but she does look after him. Karen babies the horse and fusses over the tiniest aspects of his well-being, and it is important, Lendon says, that he "has his *own* person."

Because Anne doesn't usually control ownership of her horses, it is not often she can hold on to a horse for teaching purposes. The horse will be sold, usually to an amateur to compete over less difficult courses. This was the case with Top Seed, who for several years before he peaked was consistently in the big money. He was sold, but he eventually returned to Market Street to carry one of Anne's clients. He was an affectionate animal, and she was as happy to have him back as she had been sad to see him go. He may have to cut back on the speed and the heights, but he is still one of Anne's "lovely" ones. Play Me could teach riders of about the same skill Top Seed can teach. Keith and I talk about less speed, lower fences, and a rider new to the middle levels of combined training. He would be too hot to be a school horse, too sensitive to teach up-downers. He hasn't the patience or fortitude. But Play Me could show a fairly accomplished rider how to get around a cross-country course.

There are plenty of riders who want to acquire the skills he can pass on. Riding is a kind of primitive engineering. You wrap your legs around an animal with ten times the mass and fifty times the strength of your own body, and if you're clever, you can make good use of this superior power. If the horse has a thing or two to show you, you become more clever.

I've often thought about the first humans to mount up and what that experience must have been. Undoubtedly the humans had been watching the horses, admiring their strength and speed and grace. And maybe there was envy. Maybe the first riders wanted some of that for themselves. I like to imagine the expressions of those first riding horses when their sides were suddenly clasped by the forked creatures who had been coaxing them closer with food.

There are not many horses that can be astonished this way now. There is Przewalski's horse of Siberia, and there are feral horses, like our own "wild" horses, animals that have wandered away from human cultivation. These animals have only the most remote experience of us. They have proliferated enough to become nuisances to ranchers the way woodchucks are to gardeners. The Bureau of Land Management periodically rounds up the younger animals and ships them to other parts of the country to be acquired by private owners for a nominal "adoption" fee. I went to one of the adoption proceedings. The vast indoor polo arena at Cornell University was crisscrossed by portable steel fences. In the pens, the horses were grouped according to age and gender. Most of them were very small and nearly as light-boned as deer. The most mature of the stallions were about the height of Lendon's Last Scene. They looked like horses in an earlier stage of evolution.

The little horses were not entirely innocent of humans. But aside from finding us a convenient and plentiful supply of food, their experiences had been mostly uncomfortable or even painful. Before they left Nevada, they were tagged with a number on a rope collar, inoculated, and some of the males castrated. They had no notion of the ambitions of the hundred or so prospective buyers that milled around the pens jotting numbers and comments. They were ignorant of their own capabilities and of the words that might describe them after they had gained knowledge of humans: "bold" and "heart," "generous," "gallant," and "game"—the big words. Many of the people who would take them home knew less about horses than the animals knew about them. It is an inexpensive way to get a horse, and many of the first-time owners were complete greenhorns.

In each pen, the horses stood together warily, and when one of the group was selected by a buyer, a wrangler stepped into the pen and simply stood between that horse and the rest of the group. The horse's fear made it easy to herd it up the center aisle between the pens to a ramp leading up into a wooden chute. The walls of the aisle were draped with blue plastic tarps, and if the horse tried to turn back, the wrangler just rapped on the tarp with a whip. It all went very quietly, but even so, when a horse found itself isolated, its fear approached hysteria.

On a platform that allowed him to stand well over the top of the chute was the head wrangler, a massive, square-shouldered man of unusual calm. After the horse was run into the chute and fought or just trembled there, he gently lowered a halter under its muzzle, drew the halter up slowly, and fastened the headpiece. A long catch rope was snapped to the halter, and as soon as the horse was released from the chute, it plunged through a short, fenced-off passage to a waiting truck or trailer. Sometimes there was confusion, and the little horse would run back toward the chute or try the steel fence that formed the corridor from the chute to the loading ramp. This was the only noise and excitement in a well-ordered procedure, and visitors clustered along the loading passage to watch. Regular announcements came over the public-address system to ask people to stand back from the passage, and for the most part these directions were obeyed. An elderly professor was one of the unruly. She kept drifting back to the steel fence, drawn closer and closer to the horses until she was standing on the lower rung of the fence with her face pushed through the top bars. The head wrangler observed this, and he would occasionally pause to send a cowboy over to pry her off the gate before the next horse came scrambling through the passage. When one of the larger animals, a blue roan, leaped up into the chute and fought to climb over the top of it, the big wrangler paused. He put his hand on the shaking neck and waited a few moments. He went slowly with the halter and let the roan fight more when it found its head encased by something foreign. Then he looked over at the small person clinging to the gate and said levelly, "Lady, I can't let him out of this chute until you come away from that gate. You could get hurt—I mean he don't know anything yet."

By the time most young horses are first mounted, they have acquired far more trust of humans than these feral horses. If they're handled thoughtfully, they progress through physical sensation toward an understanding of what a rider will feel like. I caught a glimpse of this process on one of my last visits to Keith.

Over the three years I had known him, Keith had been making patient headway along a risky career path. He and Play Me Right had posted one of the faster cross-country rides at Fair Hill. He was in the money there, and that, combined with good placings at Millbrook and Radnor, had put him in the top 25 percent of U.S. event riders in 1994 FEI competition. He had been invited to train with USET coach Mark Phillips in sessions for prospective team members, and although he refused to let me make much of this, he found the sessions valuable.

Other indicators were up as well. Keith's list of clients had grown. The past winter, a time when he used to lose students and rides, had been a lot easier. His reputation has grown locally, and on days when the weather was nasty and he was faced with the prospect of no work, he would make a couple of phone calls to let clients know he could have time for them. When I mentioned that the quality of the horses he was working with seemed higher, he nodded and seemed to take satisfaction in this. He still had to freelance, and he talked hopefully about establishing a place of his own. But his foothold in the West Chester suburbs was much surer now.

One June day, we had been making his rounds of stables there, and we turned into a drive just off the Pennsylvania Turnpike. The drive eventually led to a recently built house with a small stable behind it. Keith parked the Saab and checked his watch to be sure he was on time.

"What are we going to do here?"

"I'm supposed to back a colt." He carried his saddle and helmet toward the stable, and before he was inside, he was greeted by two women. One was the owner of the colt, the other was her dressage instructor. The owner was smiling and nervous.

"I don't know. You may not want to try this today. I turned him out a little while ago, and he just tore around. He really got worked up."

"Yeah?" Keith walked back through the stable to a stall where the candidate, a three-year-old Thoroughbred, waited with his sides heaving, recovering from his raucous play. Sweat had begun to break through and trickle down the colt's neck and flanks. Even though he had worn a bridle and was familiar with it, his eyes rolled when the instructor approached with it in her hand.

"Let's see what he's like," Keith suggested. "Maybe he's got it all out of his system."

Ever since we had arrived, the women had been talking, sometimes both of them demanding Keith's attention at the same time. Keith makes his living by being polite to people, mainly to middle-aged women, so he listened and nodded and kept going with the colt's tack. Many horse people are great talkers, full of histories and opinions and affections, but this was talking of a different order, nervous filling of the time that passed as the bridle was adjusted and Keith laid the saddle on the colt's back and gradually snugged the girth. The colt actually seemed to settle a little once the saddle was secure. The instructor snapped a lunge line, a long web rein, to the bit and led the colt up a hill to where a sand rectangle leveled off the crown of the hill. Keith followed, observing the colt in an offhand way. When they arrived at the riding area, a tractor-trailer truck passed at a distance of fifty yards. The Pennsylvania Turnpike ran along the edge of the property. Traffic was steady, and on the hill there was a constant, strong Dopplerized rush punctuated by the thunder of trucks and the slamming of their loads. The colt didn't flick an ear at the noise. He had grown up there.

He looked much as I pictured Play Me Right at three

years, dark bay, nearly black, and weedy. His chest was still narrow and undeveloped, his body not yet deep enough for the length of the stringy legs, his ears and round eyes still large for his head. The instructor let the lunge rein out and the colt walked around her in a broad circle. Keith watched, nodding in polite intervals at a long account the owner launched into of the accomplishments of the colt's mother and father. When the colt came to a halt at the end of the long rein and Keith walked away from us to stand next to him, the owner continued the complicated saga for me, glancing often in the direction of her colt.

With the instructor standing at the colt's head, Keith draped an arm across the saddle, hugged the colt there, and thumped his sides lackadaisically. At the first pressure, the colt lifted its head but then relaxed and stood quietly. Keith moved to the other side to hug and thump. The colt's tail switched occasionally at flies. The woman who owned him was still talking—the mare's speed, her jump, the cross-country at Loudoun. Keith began to pop up and down beside the colt, springing off the balls of his feet with his hands on the saddle. Then to the other side to pop up and down, and next to having the instructor boost him from the knee so he could drape across the saddle on his belly. The colt endured Keith's weight stolidly. After Keith had been going through all the motions for fifteen or twenty minutes, he got a boost up to actually swing a leg over and straddle the colt. The colt tensed, lifting its head, dropping its back, and tightening all over. But this lasted only a couple of seconds. Passing trucks rumbled and clanked. Suddenly the owner was silent.

"Lead him," Keith suggested, and the colt carried its rider forward. Although his head and neck were in a low, relaxed attitude, the young horse's flickering ears showed his uncertainty. But after a minute or so, he became accustomed to moving around with the weight of a human. The woman beside me resumed the tale she had broken off, and in a moment the colt was walking in a circle on the lunge rein, beyond the immediate control of the instructor. His gait was tentative, and every few strides his shoulder dipped back shyly. Keith finished the session by riding at a walk off the line for a few minutes. Then with the instructor back at the colt's head, he slid down carefully from the saddle. As we walked back to the stable, the owner abruptly ended the narrative that had provided nervous accompaniment to the whole process. The colt had now been backed, and she said she could see that the horse would be every bit as wonderful as his mother. Keith was to return the next day and get back on the colt. "That went well," he commented, and he turned the Saab into traffic and headed for the next ride.

Afterword

"This book is just a snapshot of you and how you are with your horses now," I said to Keith Taylor. He didn't like pretense or makeup on women, and he wondered if my description of his work wasn't "a little hokey." He was right that I was after something beyond flat facts. I wanted the book to say something about the tremendous potential in the athletic partnership between humans and horses. The potential of two of the partnerships in the book has been realized since its original publication, and that of one thwarted.

Anne Kursinski and Eros mastered the fences in eight rigorous qualifying competitions to win a place on the 1996 U.S. Olympic team. That team, jumping against the daunting power of the European horses and the experience of their riders, won the silver medal. For an extended period after the games, Eros carried on his work privately. Anne's strategy had always been to "save him for the big things," and she was conserving his mind and physical resources for the 1998 World Equestrian Games in Rome. Eros topped the qualifying rounds to lead the U.S. team. But he was badly off his game in Rome, and for one long night he was in sick bay. Then—true to the show-jumping axiom that says you're either on top of the world or at the bottom of the heap— Eros flew from Rome to Monterrey, Mexico, where he won the second largest purse in show jumping and made it clear that his gifts remain extravagant.

At twelve, Eros has both the youth and the experience to look forward to another Olympic Games, and with that competition now in her sights, Anne is already back-timing their preparation to qualify for the Olympics. Against a background of legal squabbling and rancor over the selection process for the team, official interest in maintaining objectivity has made the qualification trials "very grueling and very intense" for the horse, according to Anne. Now under much discussion, the trials could become even more demanding. But, she said, no matter what form they eventually take, when the time to select the team rolls around, "Just tell me what I need to do, and I'll be there."

Lendon Gray took a strategy for Last Scene as conservative as the one Anne took for Eros. "He hacks. I mean I rarely, *rarely*, school him." Now eighteen, the little gray is better known to the public than any of the "big" horses who can outscore him in competition. As the horse confirmed his understanding of the grand prix movements, Lendon showed him occasionally, and he won often. But his great success has been his demonstration rides. He went to the 1996 Olympics not to compete but, along with Sonny's Cash Lander, to show the crowds how the moves should be made. Like Lendon herself, he has done a great deal to tell Americans what dressage is.

Lendon's single-minded, sometimes contrarian devotion to the individual horse and dressage has attracted a talented, steady crew of working students, opportunities to speak and write, and a gradual accrual of gifted horses that began with the arrival of the young Jester. Gleneden has blossomed into a kind of household, where her students assume responsibilities for particular clients and horses. While Lendon coaches more and rides less, her own learning continues because her riding is largely reserved for the horses that need help. Like Jamboree. His great capabilities were apparent when Lendon's client bought him, but when he arrived in Bedford, Lendon realized he had been handled abusively. "He was completely shut down," she said, and this made even the most basic activity, such as turning at the walk, difficult. "But I made up my mind I was not going to force this horse to do anything. I was going to show him this is something *he* can do." Jamboree responded by standing for periods as long as five minutes without moving. But eventually he got the message, and now he has the movements and the will. "It is the very, very difficult horses," Lendon said, "who have the most to give you."

Keith Taylor's progress was slower, but still progress. He took on new customers, and he was able to get out on his own, leasing a stable and living in the apartment over it. At the same time, there were setbacks. In 1996 he had a terrifying accident. He was riding an inexperienced horse on a lower-level course. The approach to the fence was sunlit. On the landing side was relative darkness. It was a light-into-dark question, and the horse misread it. He took off too early, caught his front legs on the fence, and somersaulted. The horse landed on Keith, crushing his ribs, collapsing a lung.

The accident, much to his frustration, kept him out of the big hometown action, the Radnor three-day event. But shortly after, a group of supporters sponsored the purchase of an "Olympic horse" for Keith to bring along behind Faktor. In spring 1997 Faktor launched into a string of wins at major horse trials, and he distinguished himself at Radnor, storming happily around the cross-country course. Then, during the winter, the big red horse died of a slow-working virus. Keith started over, working patiently to bring less experienced horses into advanced competition. In 1998 he returned to Radnor with Paradigm, his Olympic candidate, and a horse named McGriff. About halfway around the cross-country course McGriff made an error and chested a fence. He flipped over and landed on Keith. The accident was a mirror image of the earlier one, except that this time Keith didn't survive.

In spite of the most systematic efforts of the United States Combined Training Association, people who ride fast at big fences that don't come down put themselves at risk. It is clear to me that, even if he hadn't squared with the risks before the first accident, Keith accepted responsibility for them in its immediate aftermath. His parents had worried about where the riding was going to take him, and his mother, Milly, endured anxiety every time he headed out on cross-country. She arrived at the scene of the first crash just as the medics were putting Keith in the helicopter. He was in extreme pain. "I don't think I want to do this sport anymore," he told her. "Fine." Milly leapt at her chance. "We can talk more about graduate school." But by the time he was rolled into the hospital, Keith had recovered his purpose and was scheming about how to keep the horses fit for Radnor. "You think you and Dad could take the horses to swim?"

Acknowledgments

With their unguarded cooperation, Lendon Gray, Anne Kursinski, and Keith Taylor have made this book possible. I am indebted to each of them.

Another fundamental debt is owed to Diane L. Huber. With her insightful teaching in the U.S. Pony Clubs, it was Diane who motivated this project by impressing upon me a central fact about horses and their riders: *something is going on here, something is always going on.*

The advice and perceptions of John L. Strassburger, lifelong observer of horse sports and editor of *The Chronicle of the Horse,* have been invaluable in informing my own thought. While I was fortunate enough to speak with many competitors and grooms, owners, officials, veterinarians, and members of the equestrian press, a number of people closely involved with horses offered special expertise that was particularly helpful: Marty Bauman, Frank Chapot, Trish Gilbert, Carol Hoffman, Dr. Jack Lowe, Beezie Patton, L. A. Pomeroy, Laura Rose, and Steve Stephens.

For comments on the developing manuscript, I want to thank Dean Benson, John Bailey, and Lou Robinson. The enthusiasm of my editors, Ethan Nosowsky and Jonathan Galassi of Farrar, Straus and Giroux and Lilly Golden of the Lyons Press, has been heartening. For this and for their good advice I am deeply grateful.